Urban and Landscape Perspectives

Volume 12

For further volumes:
http://www.springer.com/series/7906

Aims and Scope

Urban and Landscape Perspectives is a series which aims at nurturing theoretic reflection on the city and the territory and working out and applying methods and techniques for improving our physical and social landscapes.

The main issue in the series is developed around the projectual dimension, with the objective of visualising both the city and the territory from a particular viewpoint, which singles out the territorial dimension as the city's space of communication and negotiation.

The series will face emerging problems that characterise the dynamics of city development, like the new, fresh relations between urban societies and physical space, the right to the city, urban equity, the project for the physical city as a means to reveal civitas, signs of new social cohesiveness, the sense of contemporary public space and the sustainability of urban development.

Concerned with advancing theories on the city, the series resolves to welcome articles that feature a pluralism of disciplinary contributions studying formal and informal practices on the project for the city and seeking conceptual and operative categories capable of understanding and facing the problems inherent in the profound transformations of contemporary urban landscapes.

Claudia Basta • Stefano Moroni
Editors

Ethics, Design and Planning of the Built Environment

Springer

Editors
Claudia Basta
Land Use Planning Group
Environmental Sciences Department
Wageningen University and Research Centre
Wageningen, The Netherlands

Stefano Moroni
Department of Architecture and Urban Studies
Milan Politecnico
Milan, Italy

ISBN 978-94-007-9585-3 ISBN 978-94-007-5246-7 (eBook)
DOI 10.1007/978-94-007-5246-7
Springer Dordrecht Heidelberg New York London

Softcover re-print of the Hardcover 1st edition 2013

Printed on acid-free paper

Springer is part of Springer Science+Business Media (www.springer.com)

Contents

Preface: Shared Spaces. Shared Values?

Claudia Basta

Nothing can have as its destination anything other than its origin

(Simone Weil, 1942)

This collection of contributions on the ethics, design and planning of the built environment is the fruit of an initially less ambitious plan. In June 2010, we brought together members of the faculties of philosophy and architecture of the Delft University of Technology in a seminar whose scope was to identify and discuss shared areas of investigation. Little did we suspect that this plan would produce so many varied and significant contributions. This was quite beyond our expectations. Given the interfaculty character of our seminar, the discussions raised clearly injected many thought-provoking ideas into the debate on the ethics of the built environment. Consequently, we decided to invite some of the most prominent international authors in the field to join in with our inquiry. What gradually emerged was an organic set of contributions that revealed consistent trajectories of ethical investigation. By publishing this volume, we hope that such lines of inquiry will strike a chord in the reader and spark a genuine interest to explore further.

From the very outset, there was wide consensus at interdisciplinary level on the urgency of incorporating ethical considerations into architecture and spatial design and planning. Throughout history, *shared spaces* have always entailed, or at least always called for, *shared values*. What makes the current moment in time particularly delicate is that the city as an entity is in transition toward a multicultural reality, and it struggles to balance historical heritage and environmental patrimony with urban and rural models receptive to the challenge of sustainable development. From the perspectives of the disciplinary outlooks involved in this volume, what makes the current moment in time even more complex is the lasting polarization of entrenched dichotomies such as the "analytical vs. the normative," the "quantitative vs. the qualitative," and – perhaps most significantly – the "neutral vs. the values-driven." To our mind, it is precisely these polarities which offer the most fertile terrain on which to implant and develop our discourse.

The Ideal City as a Convoy of Reference Values

As in other areas of public life, current debates in architectural, urban, and environmental planning gravitate around the matter of balancing tradition with cultural and technological innovation. In itself, "innovation" is often perceived to clash with the very meaning of "harmony". Our increasingly multicultural societies and multi-faceted spatial settings have progressively lost the classic connotation of "according to the rules of nature" and evolved toward that of "mirroring the reality of complexity." This is something that the architect and the planner (or simply the "designer") are called upon to consider in their theoretical as well as practical elaborations. Such complexity does not only refer to the interconnection among tangible elements of the built environment, but also to the increased relevance of their "intangible" interconnection: indeed, it is perhaps *the real vs. the virtual* that emerges as the most prominent distinction of our time. A steadily growing portion of individuals "act" and "move" daily in virtual spaces, often limiting interaction within their material spaces as result. By doing so, they prompt inevitable changes in both private and collective spatial settings. The capacity to connect with and remotely activate other people and artifacts, without actually covering material distances, has entailed a series of era-defining transformations: the home workstation has made the office redundant; the computer screen has become a high-speed freeway; downloads substitute visiting the local library; idem the doctor's surgery. Inexorably, this expanding immateriality is "devouring" materiality, or at least reducing it to a realm equated with functions performed by earlier generations. For some, this is a welcome product of technological innovation, which, like any other form of innovation, redefines the boundaries between the "individual space" and the "shared space" in which the "I" encounters the "Other."

Throughout history, the complexity of this zone of shared interaction has preoccupied thinkers of all disciplines, and not least those involved in actually designing such spaces. Faced with today's growing complexity and technological innovation, along with the challenge of sustainability in increasingly multicultural societies, we cannot avoid seriously pondering how the ideal built environment of tomorrow *could* and *should* be. This is a wholly legitimate and urgent question that all interested parties must pose themselves. Our preliminary reply is that the approach to the built environment of tomorrow requires abandoning the ambition of a generally valid spatial model in favor of generally valid ethical approaches to its conception. In essence, we believe that the disciplinary transition to activate is not outward – but somehow, inward; the answer lies in how the designer conceives the material world as *shared space*, a physical arena in which the individual encounters the Other, and in which all of society's diversities may find a common ground for construction.

One might argue that this has always been the challenge. The search for ideality in architectural and urban design has plagued intellectuals since Plato's first speculations on what constituted the "ideal city" and on how to transpose the immaterial nature of an ideal society onto a material, manmade space. Since then, it has become natural to reflect on the relation between *desired society* and *designed reality*. Two valid examples are the ideal city of the Renaissance, which celebrated the emerging civic order born from the new shift toward rationality, and the utopian cities dear to twentieth-century intellectuals, such as Le Corbusier's *Cité Radieuse*. In the case of the first, the ideal city physically embodied the new balance between the religious and political powers by establishing the urban space as the new center of daily life, fostering the "civic" identity; this shift was organic to the times and had no single author as such. Conversely, the *Cité Radieuse* was effectively the outcome of a single intellectual proposing innovation upon the society of his time; however, it also reflected a new form of centrality, namely, the role of industry and the "working man" within the city, affirming his status of "equality" and the standardization this entailed.

These two examples are like the extremes of a swinging pendulum. At the one extreme, we find an intellectual consensus consistent with the spirit of the time, by which the ideal city was fashioned by the pillars of political and temporal powers; at the other extreme, we have a thinker whose notion of ideality related to the form given to space by the driving factors of modernity, that is, the new standardized productive capacity and the social class emerging through it. In this sense, the former approach envisioned the ideal space as the static setting within which societal dynamics ought to occur and within which the resulting design must be confined; conversely, the latter approach envisioned the ideal space as the dynamic arena resulting from the new and disruptive societal relations. As such, the former city idealized differences and put the citizen in a submitted hierarchy, on top of which the designer could be interpreted as the *longa manu* of ruling powers; the latter city instead turned the new societal differences into an opportunity for highlighting the growing needs of an emerging class, which was incorporated symbolically through explicit attention to its "industrial condition." Intrinsically, the first reflects an outward, top-down urban design imposed on society and its common spaces; the second is an expression of inward tensions from below influencing the design of the shared environment.

Notwithstanding their vastly different points of departure, both "ideal spaces" respectively embody then uncontroversial reference values of harmony and rationality, classicism and functionality, and inequality and equality. It is such values that provided the grid on which the conceptions of the ideal space were drawn, in accord with their historic moment. These reference values had not only an aesthetic or functional relevance, but they were also *normative points of reference*. They expressed what the "good" space, and the "good" man acting within it, ought to be. The ideal city of the Renaissance stigmatized the new enlightened centrality of the "urban man" in history; five centuries later, by moving "the center" from cities to industrial poles, the *Cité Radieuse* emphasized the centrality of the "working man" within it.

Notably, what the two ideal cities have in common is that both expressed the values they referred to with equal strength and precision. Furthermore, they remain enduring points of reference in the handbooks of history of design, and ultimately both conceptions have influenced concrete spatial interventions ever since.

The Ideal Space in a Complex Reality

One of the crucial questions we must ask is whether it is possible to translate clear and uncontroversial reference values into architectural and spatial design in our time. And if so, is it actually desirable to "design the values of our time" and hence construct the built environment accordingly?

To approach this fascinating question, we should first take a look at certain distinctions. First of all, the work of the architect and urban planner entails rather different albeit complementary ethical implications. Put very simply, the architect deals primarily with the matter of balancing the requirements for safety, accessibility, and functionality in his artifacts with his personal aesthetic inclinations. Somehow, any architectural artifact should reflect the integration of a design "*of* the I" with a design "*for* the Other." Differently, the planner deals with the matter of formulating the policy, regulatory, and spatial frameworks within which such artifacts can be located, constructed, and accessed. While the former is therefore responsible for the more tangible and "discrete" elements of the built environment, the latter is called upon to take responsibility for its more intangible, but equally determinant, rationale.

Notwithstanding these differences, both professions have fundamental responsibilities toward the society in which they operate, and that is, primarily, a *moral* responsibility. Anti-aesthetic and functionally questionable interventions (typical of low-income residential districts, degraded historical areas, or disused industrial sites) can affect the lives of citizens to the point of provoking a sense of stigma and marginalization. Land-use plans that favor the interests of a privileged few at the expense of the rights of the many merely aggravate existing forms of inequality and make them increasingly difficult to eradicate.

So what happens when both professional figures ignore, underestimate, or disregard the implications of their work and the responsibility toward the Other that their work entails? Put simply, their infringement of fundamental moral values is given form and matter in the living environment they help to create. Sadly, this does not only relate to the complex European architectural and urban heritage; examples of urban ghettoization, architectural monstrosities, and controversial siting of impacting technologies are continually documented all around the world in both scientific and professional literature.

To come back to the opening question of this Introduction, the current challenge of architecture and planning is hence no longer that of identifying the ideal space of tomorrow but rather that of ensuring that the ethical considerations that safeguard the shared values of today are embedded in the conception of the built environment and recognizable during all processes of its realization.

While this is not the place to elaborate on the specifics of the values that *ought to be* embedded into the practice of conception of our shared spaces, the issue indisputably concerns the nonnegotiable values that liberal democracies have conquered over time. The growing interest among designers and ethicists in the "values-in-design" goes beyond the classic "form vs. function" distinction; it supersedes the mere aesthetics debate in architecture and the long-lasting debate on individual *vs.*

collective interests in spatial planning. The matter of values in design is far broader and far more fundamental and concerns the moral implications of artifacts and spatial plans on citizens and the responsibilities of the designer toward them. It involves the way the interventions of the designer enable the public to share experiences in the shared space. Nowadays, there is wide consensus that urban space must mirror values such as freedom, equality and participation, and more generally the value of intra- and intergenerational justice.

Rather than be a mere exercise for a few scholars, such values should be intrinsic to professional activities. To some extent, one might argue that any artifact or form of spatial organization tends to pivot on some guiding moral principles. The problem is the significant variance in awareness of the designer or planner regarding the clarity through which he or she translates such guiding principles into their work. Take, for example, a national monument. To contemporary eyes, the moral significance of a national monument no longer has the connotation it once held for the populations of the European cities under the architectural dictatorships of the first half of the past century. However, in their contexts, such national monuments were intended as celebrations of what was considered the *good* and *just* society at the time. Nowadays, their original moral significance is often merely commemorative, at times proudly symbolic, and sometimes highly controversial. Nevertheless, for better or for worse, the values that they once represented were and still are clearly recognizable today, together with the privileged role of the built environment in conveying them.

The more controversial translation into design and planning of the values of today lies in the greater freedom we now have in assuming even strongly conflicting perspectives. After all, true democracy entails the freedom of diversity. But while this might suggest that an ethical debate within the design and planning field is less pressing than before, the evidence is that it has never been so urgent. Never before was it so important to press the designer to strive for greater clarity and consistency when making choices which may impinge on the values of others; and it is precisely because those choices are open to misinterpretation that their significance and import requires uncontroversial identification and transparency.

As noted above, the first step in this direction does not consist of arguing over the specifics of such values, but rather of reflecting with ethical rigor on the consistency between the (intentional or otherwise) objectives of architects and planners on the one hand and the perceptions and experiences of the "users" of their interventions on the other. In this sense, our view is that any ethical discussion should foster an open dialogue between designers and this "Other," in order to *explicitly* demonstrate what the built environment often conveys and expresses *implicitly*. The ethical debate must involve designers and planners in shaping value-sensitive design processes. Furthermore, the active participation of architects and planners in this interdisciplinary exchange is a priceless opportunity for steering the ethical inquiry toward the most needed directions.

A concerted interactive exchange among disciplinary domains and practices is the core idea of the contributions collected in this volume, which have preserved the spirit of the original seminar. The contributions that follow hinge on three main areas of investigation, namely,

1. The identification of the often *implicit* values informing the process of design of the built environment and converting them into *explicit* objects of design and planning from the early stages of conception
2. The identification of the legitimate boundary between the designer's aesthetical choices and the requirements of artifacts in relation to values such as privacy, accessibility, safety, and equality
3. Assigning clearly identifiable responsibilities to the actors involved in the conception, construction, use and interpretation of the built environment, a process that involves all interested parties, from the designers and planners to citizens

Reinforcing the dialogue among the ethics, design, and planning of the built environment and charting the trajectory toward the new shared spaces and shared values between them is the ultimate scope of these lines of inquiry. Hence, more than providing any definitive replies, this volume hopes to lead to the identification of future, and equally shared, questions.

Contributors

Ernest R. Alexander APD-alexander planning & design, Tel-Aviv, Israel

Carol D. Barrett Community Development Department, City of Burbank, CA, USA

Claudia Basta Land Use Planning Group, Environmental Sciences Department, Wageningen University and Research Centre, Wageningen, The Netherlands

Timothy Beatley School of Architecture, University of Virginia, Charlottesville, VA, USA

Marc Bekoff Ecology and Evolutionary Biology, University of Colorado, Boulder, CO, USA

Francesco Chiodelli Department of Architecture and Urban Studies, Milan Politecnico, Milan, Italy

Thomas L. Harper Faculty of Environmental Design, University of Calgary, Calgary, AB, Canada

Stefan Koller Faculty of Technology, Policy and Management, Section Philosophy/3TU. Centre for Ethics and Technology, Delft University of Technology, Delft, The Netherlands

Stefano Moroni Department of Architecture and Urban Studies, Milan Politecnico, Milan, Italy

Piet Rietveld Faculty of Economics, VU University of Amsterdam, Amsterdam, The Netherlands

Sabine Roeser Faculty of Technology, Policy and Management, Section Philosophy/3TU. Centre for Ethics and Technology, Delft University of Technology, Delft, The Netherlands

Lara Schrijver Faculty of Architecture, Delft University of Technology, Delft, The Netherlands

Stanley M. Stein Faculty of Environmental Design, University of Calgary, Calgary, AB, Canada

Emily Talen School of Sustainability and Geographical Sciences and Urban Planning, Arizona State University, Tempe, AZ, USA

Bert van Wee Faculty of Technology, Policy and Management, Section Transport & Logistic, Delft University of Technology, Delft, The Netherlands

Jeroen van den Hoven Faculty of Technology, Policy and Management, Section Philosophy/3TU. Centre for Ethics and Technology, Delft University of Technology, Delft, The Netherlands

Part I
Rising Problems

Chapter 1
Values in Planning and Design: A Process Perspective on Ethics in Forming the Built Environment

Ernest R. Alexander

1.1 Introduction

The role of ethics in forming the environment we live in is usually addressed in terms of relationships between ethics and aesthetics in design. This serves as our point of departure, taking the architectural perspective on design of the built environment. From a planning perspective, design of the built environment implies something quite different. We begin, then, with a review of how ethics enter the intentional shaping of the physical environment from each of these two perspectives. Next, this chapter explores the causes and implications of the different approaches to ethics that are revealed. This begins by asking what is really meant by "design", when the designed subject is the formed environment that makes up our lifeworld. Answering this question involves a critical deconstruction of planning and design processes from design theory and research, and planning theory and research perspectives.

The implications of this exploration become the premises for the argument that follows. Ethics come into design and planning through the link that we discover between ethics – as prescriptive norms – and values and, thus, through the act or process of evaluation that is an integral part of design and planning processes. But analysis of these activities reveals that evaluation in design and evaluation in planning are quite different. This explains the observed differences between the various "design professions" in their ethics of forming the built environment and poses the intrinsic problem in developing and applying an integrated ethics of planning and designing the built environment.

E.R. Alexander (✉)
APD-alexander planning & design, Tel-Aviv, Israel
e-mail: eralex@inter.net.il

C. Basta and S. Moroni (eds.), *Ethics, Design and Planning of the Built Environment*, Urban and Landscape Perspectives 12, DOI 10.1007/978-94-007-5246-7_1,

1.2 Ethics in Design: Past and Present

1.2.1 Designing the Built Environment

For architecture – as a profession and a discipline – the built environment is the product of design, from the design of individual buildings through built complexes and ultimately to whole cities. From this perspective, there are three ways in which ethics relate to design. One is the link between ethics and aesthetics, another is how ethics relate to function, and the third is the recent concern with ethics of the environment.

Ethics and Aesthetics. Architecture's concern with aesthetics goes back to antiquity. In his textbook *De Architectura*, Vitruvius cites three guiding principles: *firmitas, utilitas and venustas*[1]; today, the last of these means beauty, and it is the heading under which Vitruvius invokes the aesthetic norms of classic design.[2]

Here ethics enter design through aesthetics, which this approach considers identical. There are various arguments for this identity. One is their common base: ethics and aesthetics are both concerned with values (Collinson 1985)[3]; another suggests the appreciation of art as a form of moral engagement, implying a relationship in art between aesthetic and moral values (Dean 2002). Finally, architectural expression of values through its aesthetic dimension can influence social life and behaviour (Rondanini 1981), suggesting aesthetics as the channel through which design introduces ethics in forming the built environment.

Alberti's Renaissance treatise *De Re Aedificatoria* has a different structure from Vitruvius' and treats design aesthetics in another fashion. Of its three "levels of human activity" – *necessitas, commoditas, voluptas* (necessity, commodity, aesthetic pleasure) – two include aesthetic principles. Under the last, beauty invokes an aesthetic focused on orders, ornamentation and proportion, linked to ethics through its prescriptions of economy and prudence. *Commoditas*, too, involves aesthestics of another kind under its subsumed human aspect of "desire", to generate a "theory of program": a set of universal design rules that also covers the planning and construction of the (ideal) city. Based on the principle of fulfilling desire, these rules are intended to be flexible and adaptive to meet specific users' wants and clients' preferences, thus basing Alberti's design aesthetics on an explicit ethics of responsibility (Choay 1997, pp. 67–81). Unlike Vitruvius' treatise where the identity between ethics and aesthetics is implied, in Alberti's, the link between aesthetic prescription and ethical principles is explicit.

Ethics and Function. Both Vitruvius and Alberti address function in their treatises on architectural design: Vitruvius makes functionality an explicit principle under the term *utilitas*, while Alberti covers the functional aspects of design under his headings of *necessitas* and *commoditas*. The ethical dimension of functionalism – designers' responsibility to the users of their products – was recognised in the past, as these texts show. But in traditional architecture and design, the functional aspect was secondary, overshadowed by aesthetic values.

From antiquity to the nineteenth century, in the education and socialisation of the architect-designer, the artistic-aesthetic and individual-expressive dimensions of design were much more prominent than its social, practical and functional aspects. With some qualifications (discussed below), this orientation has persisted to the present. An extreme reflection of this image is Ayn Rand's (1943) character, the architect Howard Roark, whose aesthetic glorifies the libertarian value of unbridled individual creativity.

The twentieth century saw the countermovement to this orientation, with the rise of international modernism. This school's doctrine, as propounded by its most eloquent exponent, conferred aesthetic primacy to function: "form follows function" (Le Corbusier 1923).[4] In practice, Le Corbusier's and other modernist designers' functionalism was more symbolic than real, and their sense of ethic accountability to their user clients was still rather limited.[5]

A subset of this architectural movement emphasised a design ethic of social responsibility influenced by Marxist conceptions of social justice (Wurster 1965). One member of this group was the Bauhaus school at Dessau under Walter Gropius' leadership; another was the Dutch De Stijl movement and its main architectural exponent, Gerrit Rietveld. Prominent products reflecting this orientation were Walter Gropius' workers' housing in Dessau, Germany, Gerrit Rietveld's workers' housing estate in De Hoek, the Netherlands, and the German Weissenhof Estate directed by Mies van der Rohe,[6] but perhaps its best manifestation is the Karl Marx Hof in Vienna.[7]

Ethics and Environment. The principles of traditional architecture prescribe design that is compatible with its environment. For European classical and post-classical architecture, Alberti's treatise sets out rules for siting buildings and settlements, and laying out complexes and cities, that will ensure their effective performance in their particular location and climate and their harmonious integration into their natural surroundings. In another chapter, it ensures conformity of new designs with their socially formed visual environment by its rules on architectural styles and orders (Choay 1997, pp. 73–98). In traditional Chinese architecture and urban design, the principles of feng shui[8] ensured compatibility of buildings and settlements with their particular natural environments.

Emerging modernising trends in early twentieth century architecture expressed their environmental ethic through the design concept of contextualism: "the goal of contextual architecture is to preserve the natural beauty of the ... site ... through careful design that relates to its surroundings ..." (Wolford 2004, p. 151). Contextualism was embedded in several architectural styles and projects in the first half of the last century, including California regional architects of the 1920s, Frank Lloyd Wright and his followers, Alvar Aalto and Kenzo Tange and his Japanese school (Wolford 2004, pp. 64–146).

Later in the twentieth century, contextualism saw a revival in the UK and the USA. In Britain Prince Charles articulated its doctrine in several attacks on conventional modernism in the late 1980s, and in an exemplary plan for Poundbury new town on his Cornwall property for which he commissioned architect-planners Leon Krier and Andres Duany (Williams 2004, pp. 43–52). The New Urbanism movement in the USA expressed its version of contextualism that originated in the earlier design doctrines of Rob and Leon Krier and the Italian morphologists in neo-traditional planning of its co-founders' exemplary projects on the Atlantic coast, for example, Seaside, FL (Ellin 1999, pp. 100–110; Howard 2002). The New Urbanism's contextualism, manifested in its revival of premodern architectural forms and traditional design ideals (Hirt 2009), is becoming increasingly widespread in the USA and Europe.

Integrating Ethics into Design and Planning. Recent awareness of ethical issues concerning the built environment has produced some interesting efforts to develop or present an ethics as a prescriptive model for design. One is Warwick Fox's "integrated-comprehensive ethic" (Fox 2006) that claims to fill the gap left by conventional ethics, which fail to address the human-constructed environment. Consequently, Fox makes this the third dimension of his complex multidimensional model, where the other two dimensions are interhuman ethics and ethics of the natural environment, introducing design considerations into the realm of ethics, as ethics that relate to relatively intangible concerns,[9] "specifically, (buildings') degree of contextual fit" (p. 48).

Another such effort is Talen and Ellis' (2002) presentation of the New Urbanism as a systematic ethical model for designing the form of the built environment. The principles of the New Urbanism are offered as a normative theory of urban design, a substantive ethics for planning as opposed to the current focus on procedural issues and postmodern relativism. The design professions need "a strong well-articulated theory of good city form" (p. 38) to pursue their "quest for excellence, quality and beauty in our built environments" (p. 39). These principles are the best available current model to produce "good urban form" and enable the integration of urban design theory into the array of current prescriptive planning theories (Talen and Ellis 2002, pp. 39–41).

1.2.2 Forming the Lived-in Environment

In planning, design is also involved in forming the built environment, but not as in architecture or urban design. "Planning … is the design of a course of future action to reach (desired) ends … Urban planning is the collective management of urban development, us(ing) purposeful deliberation to give shape to human settlements" (Fischler 2012, p. 108). The ways in which planning involves ethics in forming the built environment are also different from what we have observed above in architectural and urban design theory and practice.

Cities and human settlements have been planned since the dawn of time (Alexander 1992, pp. 15–39), but for our discussion it is useful to distinguish between traditional and modern planning. Here modern planning is defined as a conscious and reflexive practice (distinct from architecture and urban design) that is associated with the appearance of urban planning as a recognisable profession and discipline, emerging in the late nineteenth century and maturing in the twentieth century.

Traditional planning ethics were its expression of collective values: the commonly held social and community norms that influenced (and sometimes even determined) the form of vernacular settlements. Such planning has existed for almost as long as societies have. In early civilizations too, the built environment was not limited to its planned elements (fortifications, palace and temple complexes) but was also controlled by formal codes and informal norms of behaviour to conform to an overall conceptual scheme. Sometimes this scheme reflected an "ideal plan" that was current in the respective society, for example, Chinese, Japanese and Mughal capitals; in other cultures, for example, Khmer Cambodia and pre-Columbian America, it was a more schematic ordering system.

Traditionally planned cities flourished in classic European civilizations: Hippodamus' Miletian grid and the Roman castrum plan were models for planned systems of spatio-political order, and Roman building codes and city regulations are well documented. In medieval Europe complex systems of laws and codes controlled development and construction in cities and regulated the use of the built urban environment. This was planning without designed plans, to prevent nuisances and defend property rights (Alexander et al. 2012, p. 74).

In traditional planning, values form the built environment through the society's institutions, which exist to effectuate commonly held values through collective action and individual behaviour (Alexander et al. 2012, p. 75). Thus, societies' formal institutions – religion, tribal organisation, theocratic regimes, autocratic regimes and democratic regimes – inculcate norms, promulgate codes and enact the laws that make up the "toolkit" of traditional planning. This holds true as well for most planning today, in which formal social institutions and informal cultural norms dominate formal planning actors.

Modern planning has suffered from an identity crisis almost as long as it has existed, and defining it conclusively is difficult if not impossible (Alexander 2005a, pp. 93–98).[10] For the purpose of this discussion, modern planning is identified as a professional practice and its associated disciplines variously known as spatial planning, urban and regional planning, *planologie*, *urbanismo*, *amenagement de territoire*, etc. Ethics are involved in spatial planning[11] in two ways. One is through planning practioners' professional ethics; the other is through the procedural-participatory aspect of planning.

Planning Ethics and the Public Interest. Formally, planners' professional codes include three dimensions: their accountability towards professional colleagues, their responsibilities towards their formal clients and their social-ethical obligations. Our discussion concerns the last, which professional codes express as planners' responsibility to act in the public interest.

Though the principle of the public interest goes back to antiquity, more recently its utility and even its substantive existence have been contested. Nevertheless, its value for guiding collective decisions and action has been reasserted, especially for planning. Modern planning uses the public interest as a legitimating principle and as an ethical norm to guide and evaluate proposed action.

The public interest includes ethics in its various concepts and applications. Substantive concepts include consequential ethics as utilitarianism, widely applied in planning practice and evaluation (e.g. benefit-cost analysis); unitary approaches such as communitarianism and etatism that are applied through political, administrative and judicial review and decision-making; and deontic ethics associated with individual and collective rights, applied through the rule of law in administrative deliberation and judicial review. The procedural orientation also includes deontic ethics such as due process and the dialogical principle as expressed in political discourse and democratic participation (Alexander 2002, pp. 226–237).

Democratic Participation: The Procedural Ethics of Planning. The salience of the value of democratic participation in planners' prescriptions for forming the physical environment is expressed well in Fischler's "theses" (2012, pp. 110–111):

> (G)ood urban planning … help(s) to make planning an opportunity for public learning and public deliberation … compensat(ing) for inbalances of power in society in terms of access to information … decision-making forums … and to decision makers … work(ing) to make public discussions … transparent, constructive, and respectful of differences. Good plans benefit from the input of all people and institutions … (and) stakeholder participation at an early stage in the planning process.

This procedural side of planning distinguishes it from planning-as-urban design, manifest in "the tension … between urban planning as the design of spaces and places and urban planning as the design of institutions and processes" (Fischler 2012, p. 108). It also accounts for the lack of convergence between the ethics involved in the two forms of design. The different ways in which we understand the term "design", how and why these affect the ethics of forming the designed environment, are addressed next.

1.3 Design in Designing Places and Planning Environments[12]

Usually, design means creating form and is associated with the arts and the "design professions". Architects design buildings; engineers design highways, bridges and machines; product designers create appliances and their packaging; planners shape cities; and urban designers form their streets and squares. But a more nuanced conception of design raises definitional dilemmas.

1.3.1 Models of the Design Process

In its conventional meaning, design is regarded as a particular process or activity, which is invoked to address issues or problems that have design solutions. An alternative definition, however, sees design as a universal part of all decision-making or problem-solving behaviour. In this view, design is a subset of a more extended and complex process (not necessarily reflexive or linear), something that every actor[13] does (intuitively, informally or deliberately) in developing and considering alternative courses of action. From this perspective, the design process does not produce the final product that is implemented, but rather generates the alternatives that are the objects of evaluation and choice.

These approaches to design – the first favoured by designers and researchers-scholars in the design professions and the second implicit in planning and decision theory models of problem-solving processes – include different views of the relationship in design between creation (the actual development of alternatives), evaluation and choice. These are also implicit in the models that students and researchers of the design process have proposed over the years and which are still current today.

Model I was the first to emerge from the systematic exploration of the design process in decision-making and problem-solving. In essence this model, pioneered by Herbert Simon and subsequently maintained by his followers (Newell and Simon 1972), sees the design process as a form of search. Here design is the result of enquiry, research and analysis in a manner wholly consistent with the scientific method, distinguished only by its deontic-prescriptive logic from the scientific descriptive-explanatory mode. Simon's model of design as systematic search ignores the tension between creativity and evaluation because creativity is just finding the solution that is there and fitting it to the problem at hand by association and/or adaptation.

Model II developed as a description of the realities of design activity, based partly on the findings of empirical research and reacting to the prescriptive focus of model I. Here, design involves more than the abstract logical operations of model I; rather, it is seen as a basically non-rational process[14] that draws on deep subconscious images, associations and codes (Hillier et al. 1972).

An important difference between the two models is their approach to problem types and what kinds of problems they see as involving design. In the first, problems vary along a continuum from the novel to the routine, and though they may be complex, all are amenable to model I-type "design" solutions. Model II implies a dichotomy between routine and novel problems, associating design particularly with the latter. Here design focuses on innovation and creativity, to beg the unavoidable paradox: the intrinsic anomaly of creation (Hausman 1975). Implying the basic non-rationality of any decision-making or problem-solving process that includes creative design, this paradox limits this model's empirical analytical-explanatory potential, and its prescriptive utility approaches zero.

Model III: With this model design and planning theorists and educators have tried to produce a synthesis of the two preceding models, which acknowledges the rational and non-rational elements of design and problem-solving processes. Like the first model, it is also essentially prescriptive, but it envisages a process that includes both the logical-analytic aspects of model I and the intuitive-creative characteristics of model II.

One version of this model, the "dual processing" design process, proposed a systematic integration of creative design into the traditional deliberate-rational problem-solving framework of the design professions. Donald Schön's (1983) "frame-reflective discourse" is another more descriptive and more pragmatic version. The action-centric model is a contemporary version of model III. Claiming empirical support and drawing on Schön's reflection-in-action paradigm, this model describes design as a recursive process alternating between problem framing, making tentative design decisions and modifying them after evaluation to evolve the final design solution (Dorst and Cross 2001).

What do these models suggest about the relationship between "creation" by design and the evaluation of design proposals or alternatives in the playing out of design and planning processes? Clearly, evaluation will be different when it is applied in a relatively formal and rational decision-making process (like the design process in model I) than the kind of evaluation that is involved in the intuitive-creative design process of model II or the repeated evaluations that are integrated into the recursive-reflexive design process in model III. We will return to explore these differences after reviewing the role of design in the planning process.

1.3.2 Design and Evaluation in the Planning Process

Planning theory offers several models of planning processes, which combine in different degrees normative-prescriptive and positive-descriptive dimensions. The classic *rational planning* process – primarily linear but with some recursive elements – is essentially a normative model, in which design (of alternatives) and evaluation are distinct successive stages. Critiques of this model produced a number of modifications that reflect their authors' versions of bounded rationality.

Planning Models – Bounded Rationality: Together with his definition of design, which resulted from his critical analysis of supposedly rational decision-making processes, Simon proposed a model describing actors' bounded rationality. This distinguished decision-makers' activities from prescribed rational goal seeking by calling them *satisficing*. In satisficing, design has a different role than in the rational decision-making process: instead of searching for or developing alternatives (a process that must be unbounded by definition if it is to be sure of including the optimal course of action), the "designer's" search is limited to finding the nearest available option that satisfies a (provisional) set of choice criteria. Evaluation, too, is different here: applied successively to a single feasible option (rather than the comparative evaluation of the rational process), less extensive evaluation is appropriate, while the selected method must be capable of informing the "go-no-go" decision. Benefit-cost analysis is the perfect evaluation tool for the satisficer.

Disjointed incrementalism is another form of bounded rationality, where actors avoid deliberate design, but "muddle through" to arrive at decisions by spontaneous mutual adjustment in a sociopolitical "market" (Lindblom 1965).[15] Here, evaluation is also informal or limited, to choose between feasible alternative courses of action that promise only little change in the status quo, and benefit-cost analysis is popular.

Dialogic Planning: More radical critique of classic rationality has proposed other planning models. These see planning more as an interactive dialogical process than a rational or boundedly rational decision-making process (above). Some (e.g. *communicative practice* and *collaborative planning*) are communicative-consensual, based on Habermasian communicative rationality. Other postmodern models are conflictual, proposing planning as Foucaultian *strategic action* or as Mouffian *agonism*, while more radical postmodern approaches advocate planning as Lacanian *reflection* or Deleuzian *bricolage*.[16]

What these dialogical planning models have in common is their unconcern with evaluation as a recognised part of their planning processes. This is not surprising, given their exclusive focus on the procedural aspect of planning, while the role of evaluation in previous models related to the substantive content of plans and decisions. We can assume, then, that here evaluation is also dialogic and that ethics and values enter the planning processes these models prescribe in the same ways described above under the public interest and procedural planning.

1.3.3 Values and Evaluation in Designing and Planning Formed Environments

Our review of design and planning models focused on the implied relationships between design, planning and evaluation. First, we found two contrasting uses of the term design. One used it to encompass the whole process of giving shape – in our case to the formed human environment; the other limited the term to the stage of designing the alternatives that become the objects of evaluation and choice for elaboration and action.

The first approach, as reflected in design theory and research, basically identifies two models of design: design as a deliberate process of heuristic or systematic search and adaptation or design as an intuitive process of innovative creation. The second approach, espoused in planning and planning theory, subsumes design under several planning models that can be broadly grouped under two headings: rational planning, planning as a deliberate (rational or under bounded rationality) decision-making process, and dialogic planning, planning as a communicative process. What can this tell us about values and evaluation?

In reviewing all these design and planning models, we can find several kinds of evaluation, which differ and have different relationships to autonomous design or the "design" stage of problem-solving and planning processes.[17] They are (i) dialogic evaluation, (ii) deliberate-formal evaluation, (iii) quick or informal evaluation and (iv) intuitive-abductive evaluation.[18]

Dialogic Evaluation: In dialogic evaluation the actors involved in the planning or design process use a dialogic process to evaluate the proposed design, plan or course of action or the possible submitted alternatives. This involves formal and informal discussion and debate among the responsible actors; the arenas for this dialogue are usually formally constituted decision-making bodies: executive boards and directorates, governing commissions and committees, legislative committees and courts of law. Dialogic evaluation can take several forms, and the planning-design of a complex project often combines them. In many cases, dialogic evaluations also include one or more of the other evaluation forms (below), for example, when a formal evaluation method is applied in the course of a planning commission's administrative review of a proposed development or when a more informal benefit-cost analysis informs a budget committee's review of alternative projects.

One form of dialogic evaluation is administrative-managerial deliberation: this occurs when proposals for action are deliberated for adoption and implementation within an organisation. Such deliberative evaluation is widespread, for example, corporations' management reviewing alternative locations for a proposed R&D complex or campus, government budget committees evaluating strategic infrastructure projects for prioritisation and inclusion in the capital budget and land developers' executives evaluating proposed project plans for investment.

Another is administrative review, when a responsible decision-making body evaluates proposals that other parties submit for its required approval. Dialogic evaluation in administrative review is universal, including legislatures' committees reviewing and evaluating proposed laws and regulatory bodies' evaluation of proposals, plans, programs and projects for licensing or permission. In forming the built environment, dialogic evaluation as administrative review has an important, perhaps critical role (though underestimated in design and planning theory) when planning bodies review plans and project proposals and design committees evaluate submitted projects for approval.

Finally, there is judicial review, which involves dialogic evaluation by courts, usually of decisions arising from preceding dialogic evaluations that are appealed by affected parties. In judicial review the dialogic evaluation aspect is striking because the process essentially involves appraisal of the relevant decision or action for its conformity to valid law and its respect for the appellants' relevant rights. Judicial review offers a clear case of evaluation introducing ethics and values into the process of forming our lived-in environment. It also demonstrates the influence of one kind of ethics: the rule of law that is applied in judicial review expresses a deontic ethics – based on rights and obligations – not the consequentialist-utilitarian ethics often implicit in planning decisions and design-development choices.

Dialogic evaluation is an important channel for introducing ethics and values into planning and designing the built environment – those ethical prescriptions, procedural and substantive values that are important (consciously-reflexively or cognitively-intuitively) to the actors in the process and the institutions they represent. Thus, one dialogical evaluation might adopt a consequentialist-utilitarian ethic that reflects powerful actors' interests and their preferred substantive values: freedom of individual action and efficiency of collective undertakings. This might be in the context of an essentially incremental planning process and might include justification by simple benefit-cost analysis. Another dialogical evaluation, in the context of a formal dialogical-participative planning process, might reflect a more deontic ethic and participants' procedural and substantive values such as due process and fair distribution, sustainability and social justice.

Deliberate-Formal Evaluation: This evaluation process is formal and relatively extended, when systematic evaluation methods are applied to appraise preformed (designed or found) alternatives by judging their projected consequences and impacts on a set of preference criteria. This is evaluation as prescribed for the rational planning process, which is still extensively practised, based on the principle of substantive rationality.

In this planning process, formal evaluation follows the design stage, which produces the objects of the evaluation, and ethics (with the values on which ethical norms are based) have an important role. The first time values are identified or implied is when the goals or objectives of the undertaking are specified. These are later elaborated and operationalised in setting out the program or design brief for the design stage and influence (reflexively or intuitively) the consideration of appropriate evaluation methods. Finally, the adopted goals and objectives are translated into outcome appraisals and/or performance measures in specifying and prioritising the criteria to be applied in the evaluation stage.

The sources of the values and ethics applied in this planning process at various stages differ, depending on how the process is carried out. The critical factor is how closed-technocratic or how dialogic-participative the process is; this can also vary in each stage for any specific planning project.

In a closed-technocratic process, active involvement is limited to those individuals and social units participating in their roles as designated institutional representatives and expert professionals; they are also the ones who input their relevant values into the planning process. So, for example, in the initial goal-setting stage of a metropolitan transit planning project, the active participants might be officials of the metropolitan transit agency, elected politicians representing the metropolitan government or participating local governments, professional transportation planners leading the project planning team and other professional experts representing relevant planning agencies or concerns. The evaluation stage in such a project might involve a consultant expert evaluation team applying formal evaluation methods (e.g. combining a detailed benefit-cost analysis with a multi-criteria evaluation matrix) and developing criteria by "armchair" imputation of relevant (public and/or particular interests') values and goals.

Dialogic-participative planning involves a wider set of active participants, which can extend to the relevant public at large. This process is organised to enable the active involvement of all the relevant parties and the input of their interests and values into the planning process,[19] to ensure that the resulting policies, plans or projects reflect these values as well as possible. If the metropolitan transit planning project described above were structured as a dialogic-participative process, its evaluation stage would be quite different. The designed alternatives might be exhibited for public comment, be the subjects of directed discussion by focus groups drawn from relevant sectors and interests and be the objects of facilitator run decision-aided evaluation exercises by selected stakeholder representatives. The inputs from all these participative arenas might be aggregated and integrated in a final formal matrix evaluation developed and run by an expert team supervised by and responding to a project planning steering board made up of representatives of the responsible agencies, critical stakeholders and affected interests.

Clearly, we can expect such a planning process to reflect a different – and especially more diverse – set of values and ethical orientations than the closed-technocratic process described before. In particular, because the value inputs in a more closed professional-technocrat-dominated process are limited to a small set of powerful actors, they are likely to reflect a more instrumental and narrowly goal-oriented orientation than a dialogic process that engages a wide set of involved actors and affected publics.[20]

Quick or informal evaluation is common in planning and design processes based on bounded rationality, reflecting frequently observed models such as satisficing and incrementalism. Such evaluation is usually supported by less formal or extended analysis to compare proposals, ranging from "back-of-the-envelope" computations to show alternatives' performance on simple criteria (e.g. minimum direct costs) to simple benefit-cost analysis measuring their economic efficiency.

In relatively closed, stakeholder-controlled and professional- and technocrat-dominated planning and design processes, quick informal evaluation is much more common than appearances might suggest. That is because this kind of evaluation often informs the responsible actors' actual decisions and actions, contrary to what formal protocols show. Such documentation includes the analysis and conclusions of formal-expert evaluation of alternatives, but in-depth case study reveals that this evaluation is really only ex post rationalisation: window-dressing to justify an already finalised decision.[21] This closed process kind of planning-design is also much more widespread than it seems, because its true nature is obscured by a formal apparatus of token participation, which is often adopted to satisfy political pressures or legal mandates.[22]

Quick informal evaluation usually reflects a consequentialist-utilitarian ethics and its powerful stakeholders' procedural values of effective decision-making and efficient implementation. Such evaluations are usually in the context of goal-oriented planning-design processes, in which the prominent actors' values are manifested in their project's instrumental goals. These will vary depending on the nature of the project and in themselves may be quite unexceptionable.

Thus, a planning-design team working on a neighbourhood development for a land development corporation may be aiming to produce a marketable project to maximise its investors' profits, but its espoused goal (not necessarily incompatible with this economic objective) will be to produce a built environment that supports its residents' aspirations to a good life as they see it. An interagency review committee evaluating preliminary organisational and financing options for a national highway project may focus on privatisation alternatives,[23] motivated by neo-liberal values of shrinking government and pursuing its mandated goals of satisfying projected traffic demand while minimising public costs.

Intuitive-Abductive Evaluation: Design, as described in model II above, also includes a kind of evaluation, but it is radically different from the other kinds of evaluation presented here. In this model, design is a largely intuitive process that evokes subconscious associations and applies abductive reasoning. Here, possible designs or potential design components emerge in the designer's consciousness and are tentatively noted (usually in graphic-image form) for reflection and review for "goodness of fit" with the designer's general unformed and informal aspirations and/or other possible elements of the emerging design. The designer, in this account, may be an individual, when the design process is (at least partly) an internal dialogue. But this process can also be interpersonal and play itself out as a dialogic interaction of a design team.

Design, then, is a recursive process in which creation-forming or elaboration alternates with intuitive-abductive evaluation of tentative design options. This produces the convergence around the design that emerges by iteratively dropping from further consideration ideas or elements found inappropriate while replacing them with others that might be better. Though it is frowned upon in prescriptions for design that enable or enhance creativity and innovation, some evaluation in the course of designing is unavoidable if there is to be any convergence on a finally formed design for elaboration and implementation.

In this model of design, there are no clearly defined goals or objectives that would introduce values into the design process, though the designed product is often shaped in critical ways to conform to a design brief or program[24] that formally constrains (otherwise infinite) feasible options. Intuitive-abductive evaluation is where the designer's values come to bear in the design process where, by its nature, these values tend to reflect aesthetic considerations. It is hard to imagine a designer dropping an idea because it is inadequate in promoting social justice, though it might be eliminated because it cannot be fitted to the program requirements. It is easy to think of designers rejecting a possible design idea or element that occurred to them because it does not fit, in the formal-aesthetic sense, into the emerging overall concept or because it is formally inappropriate for integration with adopted or preferred design elements.

In design, therefore, aesthetic values are paramount,[25] and the view still prevails that ethics and aesthetics are one. Recent experience confirms this: ethical and value considerations enter design through aesthetic movements and styles. Thus, functionalism (and professed responsibility to users) is reflected through the modernist-international style, social justice through the modernist-workers movement of the 1920s–1930s, and environmental ethics and sustainability through the contextualism of the California regional school, the twentieth-century British contextualists and the contemporary New Urbanism.

1.4 Implications: Ethics and Values in Forming the Lived-in Environment

1.4.1 How Do Ethics and Values Differ in Design and Planning?

The above review of ethics in design and planning revealed significant differences in the ways that ethics and values found expression in designed environments or became manifest through planning processes.

Ethics in design takes two distinct forms. The high value attributed to beauty identifies ethics with aesthetics. The *ethics of aesthetics* is explained by their common concern with values, making aesthetics a form of moral engagement and enabling values to influence social life and behaviour through the aesthetic dimension of design. Through the ages and into the present, and across civilizations, aesthetics have been the dominant ethic of design.

The other ethic expressed in design is the *ethic of responsibility*, which has various objects. The most prominent ethic of responsibility is the designers' obligation to their users. In the normative principles of architecture and urban design, from antiquity to the present, this comes under the heading of function. This ethic is manifest as universal normative principles to ensure the utility of design products and prescriptions for adaptable design to meet the specific needs of particular users. Modern *functionalism* shows the integration of this ethic into prevailing aesthetic norms.

Another way that the ethics of responsibility show up in design is in the principle of appropriateness, or "goodness of fit". In terms of designers' responsibility to adapt their designs to their neighbours and their proximate environment, this norm blends into prescribed aesthetics under the concept of *contextualism*. More recently, this norm has been expanded to embrace the human and natural environment as a whole, leading to the adoption (by design theorists) of environmental ethics and *sustainability* as a design principle.

Ethics in planning overlap with design ethics to a limited extent, but mostly they are different. In planning there is also the *ethic of responsibility*, but little parallels the way it applies in design. This part falls under planners' professional codes of conduct, which demand their responsibility to their clients: the planner's counterpart to the designer's users.

In planning the ethic of responsibility applies in ways that are absent from the norms of design to reflect the planner's responsibility to all those affected, directly and indirectly, by planning decisions and their implementation. One way is through the *public interest*, which planners' professional codes identify as their primary objective. The identification and operationalisation of the public interest in a particular case or issue is an important way of introducing values into the process of forming the lived-in environment.

The other way is through the principle of *dialogic participation* in the planning process, which acknowledges all the involved parties as (in principle) planners and recognises them as important sources of valid information and knowledge. All current models of planning premise dialogic participation: in modern (consensual) communicative practice and collaborative planning, it is critical and explicit and implied in postmodern conflictual models. Present norms of good planning aspire to incorporate the most diverse set of values possible in the plans that result from a participative process that involves all interested and affected parties and relevant stakeholders.[26]

Summary: Two differences between ethics in design and ethics in planning are striking. One is that all the ethical concerns for design are substantive: what a good design should be; for planning the concerns are essentially procedural: what makes a good planning process.[27]

The other is the different subjects to whom the prescriptions apply. In design their direction is unambiguous: they are meant for the designer. The designer must conform to the relevant aesthetics, and the ethics of responsibility require designers to fulfil their obligations to the users of their products, to the proximate social and built environment and to the human and natural environment at large. The designer is the autonomous creator (constrained only by the limits these ethics imply) and the main source of the values that are ultimately expressed in the designed environment.

Ethics in planning are completely different: their direction is ambiguous, and the autonomy of their subject – if taken to be the professional planner – is denied. The prescribed norms involve eliciting-determining a situation-specific substantive public interest (usually in a dialogic process of political discourse) and require a dialogic-participative process that expresses the most diverse possible set of values in the resulting plans.

These ethics imply symmetry between planners and others: relevant stakeholders, interested parties and publics who are involved in the planning process and affected by its impacts. The planners' influence on the planning process and its products is limited, and their personal-professional values are relatively insignificant. In its ethics, modern planning has essentially reverted to traditional planning as described above, and the values expressed in today's planned environment (for better or for worse) are those espoused by the respective institutions and the responsible actors involved in the planning process.

1.4.2 Why Are These Ethics Different?

In looking for the sources of their approaches to the introduction of ethics and values into the formation of our lived-in environments, we discovered significant differences between design and planning. These begin with the way theorists in the respective fields model design and planning processes and conclude with their various forms of evaluation: the role of evaluation in design and planning processes and how different types of evaluation promote different ethics and values for forming the environment.

Current prescriptions for a good design process require a synthesis of systematic-methodical and intuitive-creative design approaches (model III above). But empirical research emphasises the role of intuitive design (model II above) involving subconscious associations and abductive evaluation. Abductive evaluation is also intuitive, informal and is an integral recursive-simultaneous part of the creative design process. This process may be dialogic, but the relevant discourse will be the individual designer's internal (even subconscious) dialogue or a more reflexive discussion within an interactive design team.

Here, abductive evaluation is the channel for designers' introduction of their personal-professional ethics and values into the design process and accounts for the role of these ethics in forming the designed environment. In designers' personal-professional values, aesthetics prevail, while their ethics of responsibility blend with their aesthetics to produce stylistic schools such as functionalism and contextualism.

Planning models envisage a very different process of forming the environment for human activities. Essentially, whatever form it takes, planning is a dialogic process of interaction between the responsible planners (however, these are defined) and the social-institutional environment of the relevant project or undertaking. This environment includes significant stakeholders and other involved and affected institutions, organisations, interests and publics.

In various planning models – normative and descriptive – this process (and the definition of its protagonists) takes different forms and involves different types of evaluation.

Dialogic evaluation appears as administrative deliberation for proposal adoption, administrative review for submitted project approval and judicial review. This may include formal or informal evaluation and incorporate deontic ethics: conformity to norms and rights. Formal-extended evaluation is an integral part of prescribed rational planning under bounded rationality, applying evaluation methods that aspire to substantive rationality. Applying these in open dialogic planning processes (as required by modern planning ethics), various participants introduce their diverse values; in more closed-technocratic processes (which are still common), expert evaluators impute the goals and values of important stakeholders and affected interests. Quick informal evaluation is also more frequent than admitted, in many cases informing the adoption or rejection of a proposal before formal evaluation rationalises the decision. Such evaluations often apply benefit-cost analysis supporting a utilitarian-consequentialist ethic, which reflects major actors' instrumental goals and values.

In planning, then, evaluation – whatever form it takes: dialogic, formal or informal – introduces other ethics and values than design into the final form of the lived-in environment. Rather than the ethic of aesthetics, we find the ethic of participation to elicit a diversity of social values. Rather than an aesthetic of responsibility – as functionalism or contextualism – we find an ethic of responsibility to a public interest that may take various forms, from deontic ones like planning rights and community norms to consequential-utilitarian ones reflecting involved actors' values and goals.

1.4.3 Conclusions: Is an Integrated Ethics Possible? And What Good Would It Do?

The differences between planning and design go deep, beginning with contrasting concepts of how human intentions shape the socially formed environment. The prescribed design process introduces some elements of planning in an attempt to systematise what is viewed as essentially creative design. Planning recognises design as an integral part – though small and often ignored (Alexander 1982) – of the prescribed rational planning process.

Designers – architects, landscape architects and urban designers and some engineers (when designing airports and harbours, highways, dams and bridges) – are educated and socialised to see the built environment as a product of design and to value the creative and individual aspect of design. They recognise aesthetics as the primary ethics of design, capable of integrating (as movements or styles) values such as responsibility to users and the environment. Planners are educated and socialised (with slightly less hubris) to see planning as one factor (sometimes even negligible) in the formation of the lived-in environment.[28] Current planning theory and education downplay the role of planners' personal-professional values, only promoting abstract values such as democracy, sustainability and social justice. Rather, they prescribe an interactive-dialogic planning process to effectuate the diverse substantive values of interested and affected parties.

Recent efforts to promote an integrated ethics for forming the built environment do not bridge these differences. From the design perspective, Fox's (2006) monumental attempt to integrate human with environmental ethics looks ineffective, producing a complex abstract structure that reduces in practice to an expanded contextualism. From the planning perspective, the New Urbanism's promotion of its design principles as a prescriptive ethic (Talen and Ellis 2002) looks misplaced: the attempt to supplement the procedural orientation of current planning theory and education with a specific set of substantive design-related values is (predictably) contested.

Both design ethics – aesthetics as ethics – and planning ethics – valuing dialogic participation – are valid and have their place, just as planning and design do in forming the lived-in environment. To me, combining them in some conceptual integrated structure seems infeasible, and I question whether such a structure would serve any useful purpose. The only way I see to integrate planners' and designers' ethics in forming the built environment is by institutional design (Alexander 2005b, 2006b).

This means devising, institutionalising and implementing a process in which designers and planners, professionals and participating laypersons would have assigned functions and roles enabling their constructive interaction, each contributing their skills, knowledge, talents and values to the undertaking at hand. Two factors make me pessimistic about the prospects of such an effort succeeding. One is incompatibility: the differences between designers and planners described above tend to make them incompatible for joint interaction in a team. The other is complexity: such a planning-design process would demand a structure so complex that its effectiveness would be questionable.[29] Still, to anyone who wants to try, good luck.

Notes

1. In Henry Wotton's (1624) classic translation, these are "firmness, commodity and delight".
2. Citation of subsequent sources in architectural and design theory as evidence of continuous concern with aesthetics into the present is redundant.
3. Collinson questions this identity claimed by Wittgenstein in his famous statement "Ethics and aesthetics are one", suggesting that each involves different kinds of values – ethics those concerned with human actions, while aesthetic values concern contemplation. For deeper analysis of this question, see Koeller (here).
4. This well-known phrase is actually not Le Corbusier's, but said by the earlier Chicago architect Louis Sullivan. But Le Corbusier (1923) was an explicit proponent of functionalism, for example, his statement that "the house is a machine for living in".
5. Le Corbusier's housing projects were notorious for their disregard of their residents' perceived needs, though his open plans' adaptability partly compensated for this flaw. A good example is Le Corbusier's *Quartiers Modernes Fruges*, a ~150-DU project in the Bordeaux suburb of Pessac, of which 51 houses were built: "... theoretical functionalism (was) not reflected in (Le Corbusier's) architecture, least of all in Pessac" (Boudon 1972, p. 31). The Pessac housing estate became a popular place of pilgrimage for architectural and sociological researchers (including Philip Boudon) to review and analyse the adaptative personalization of its houses.
6. Designs for the Weissenhof Estate were commissioned for the *Deutscher Werkbund* exhibition of 1927. The project consisted of 21 buildings by selected architects that included Le Corbusier, Walter Gropius, Bruno Taut and Hans Scharoun.

7. A multifamily housing complex built around a succession of internal courtyards, for about 5000 families, designed by Austrian architect-planner Karl Ehn (who was the municipality's city architect) and built between 1927 and 1930. It was bombarded and destroyed in the Austrian Civil War that preceded the Anschluss to the Nazi German *Reich*.
8. A system of geomancy originating in antiquity, to assure the "auspicious" siting and orientation of buildings and settlements, which (in addition to astrological considerations) refers to features of the local natural environment: topography, bodies of water, etc. (Chiou and Krishnamurti 1997).
9. To this reader, these appear to be largely aesthetic.
10. A complex, incidentally, that architects and design theory do not share: they have no doubts about who does design. One source of this difficulty is the conflict between planning "as a transhistorical form of social action and urban planning as a modern profession" (Fischler 2012, p. 108), which is linked with identifying its relevant actors. One approach (dominant in Anglophone discussion) identifies all the actors involved in a particular process of deliberately influencing the shape of a specific part of the humanly formed environment as planners, whether they are citizens, businesspersons, property owners, public officials, elected politicians or expert advisors in a relevant field (planning or other). The other approach (more common in continental Europe) limits the terms planning and planners to the activities and practices of socially recognised and institutionalised professionals in their relevant societal and institutional contexts. Here I adopt the second approach; proponents of the first can envisage their version of modern planning as essentially the continuation of traditional planning (as defined above) into the modern era and the present.
11. And other planning-related fields such as environmental planning, transportation planning, community development planning and regional planning, which participate in forming the physical environment.
12. This section draws on Alexander (1979, 1982, 1987). Though based on work that is more than 30 years old, a cursory review of more recent literature on this topic reveals nothing really new. 2009 sources cited in Wikipedia (2012) confirm that "no generally accepted definition of design exists", and other cited sources, including Cross et al. (1992) and Dorst and Cross (2001), continue to discuss models of design processes on the lines I present below.
13. Actors here include individuals, groups and other social units such as families, agencies and organisations. This of course raises the issue of individual vs. collective design processes, which is addressed below.
14. "Non-rational", not irrational, is intended here, but needs some explanation. The intuitive subconscious mode of design in model II does not apply the logic of conventional rationality but has a rationality of its own. This is the rationality of abductive reasoning, which designers apply rather than the deductive or inductive logic of rational analysis and decision (Cross, Dorst and Roozenburgh 1992).
15. Given this description, readers might question the inclusion of disjointed incrementalism in our discussion. But our frame of reference includes planning, and many planning theorists assert that planning is really politics. In this light disjointed incrementalism is relevant, the more so when extensive empirical research has confirmed this model's descriptive accuracy and its explanatory utility (Alexander 1992, pp. 48–49).
16. Readers who sense some dismissiveness in my account of the latter models are right.
17. The passage that follows draws on previous work (Alexander 1979, 2002, 2006a, b).
18. These kinds of evaluation are distinguished here for descriptive-analytic purposes. In reality they are not mutually exclusive; indeed, they are often mixed and one kind of evaluation may be "nested" in another (see below).
19. For a review of how this is done, that is, the various forms of public participation in planning, see Alexander (2008).
20. A recent case study analysing the development of National Policy Statements in the UK planning system and the role of evaluation instruments in strategic infrastructure and installions decisions offers striking confirmation of this hypothesis. A participatory-dialogic planning process was found to conform better to Rawlsian norms of fairness and social justice, while powerful interested actors (and deferential professional experts) preferred a more closed process

with limited participation, reflecting their instrumental values where efficient processing and rapid approval equal effective goals achievement (McKay et al. 2012).

21. An example is the case of evaluating alternative ways of funding and implementing the Trans-Israel Highway (Alexander 1998).
22. The public-participation literature offers case studies of this phenomenon that are too numerous to cite.
23. To some readers, the environmental and land-use implications of this orientation may not be evident. They include the exclusion of options such as upgrading the relevant parts of the existing network on the one hand and constructing the new highway on a right-of-way that differs radically from the existing one on the other. This is because privatisation requires operation of the new highway as a toll road, which demands a route parallel to the existing network so as to give potential users a toll-free alternative.
24. Usually such design briefs/programs focus on the functional aspects to clearly specify the functional product requirements. However, this is not always the case, and (for the purpose of our discussion) to the extent that the program identifies broader and more abstract design goals or objectives, which can be translated into evaluation criteria, the design process comes to resemble what I am calling a planning process.
25. See, for example, Roeser (here).
26. Many planning theorists (e.g. Friedmann 1982; Fainstein 2010) offer substantive norms, which planning should promote, such as democracy, sustainability and social justice, but in the final issue their prescriptions boil down to the dialogic-participative ethic presented here.
27. Though it may not be obvious, this also holds true for the public interest, which concerns the ways of determining a particular substantive public interest in a specific case – since a general substantive public interest is acknowledged to be impossible.
28. I can allow myself these observations based on personal knowledge and experience, being both a trained and practising architect and planner before my incarnation as a planning academic.
29. This has been the problem in some attempts at implementing highly dialogic-participative planning-design processes, for example, in metro-transit planning in the USA.

References

Alexander ER (1979) The design of alternatives in organizational contexts: a pilot study. Adm Sci Q 24(3):382–404

Alexander ER (1982) Design in the decision making process. Policy Sci 14(3):279–292

Alexander ER (1987) Design in planning and decision making: theory and its implications for practice. In: Alexander ER (ed) Proceedings of the 1987 conference on planning and design in urban & regional planning (1987 international congress on planning and design theory, Boston, MA, 17–20 August 1987). ASME, New York, pp 25–30

Alexander ER (1992) Approaches to planning: introducing current planning theories, concepts and issues. Gordon & Breach, Montreaux

Alexander ER (1998) Evaluation in Israeli spatial planning: theory vs. practice. In: Lichfield N, Barbanente A, Borri D, Khakee A, Prat A (eds) Evaluation in planning: facing the challenge of complexity. Kluwer, Dordrecht, pp 299–310

Alexander ER (2002) The public interest in planning: from legitimation to substantive plan evaluation. Plann Theory 1(3):226–249

Alexander ER (2005a) What do planners need to know? Identifying needed competencies, methods and skills. J Archit Plann Res 22(2):91–106

Alexander ER (2005b) Institutional transformation and planning: from institutionalization theory to institutional design. Plann Theory 4(3):209–224

Alexander ER (2006a) Evaluations and rationalities: reasoning with values in planning. In: Alexander ER (ed) Evaluation in planning: evolution and prospects. Ashgate, Aldershot, pp 39–52

Alexander ER (2006b) Institutional design for sustainable development. Town Plann Rev 77(1):1–27
Alexander ER (2008) Public participation in planning. A multidimensional model: the case of Israel. Plann Theory Pract 9(1):57–80
Alexander ER, Mazza L, Moroni S (2012) Planning without plans? Nomocracy or teleocracy for social-spatial ordering. Prog Plann 77:37–87
Boudon P (1972) Lived-in architecture: Le Corbusier's Pessac revisited. MIT Press, Cambridge, MA
Chiou S-C, Krishnamurti R (1997) Unravelling feng-shui. Environ Plann B Plann Des 24(4):549–572
Choay F (1997) The rule and the model: on the theory of architecture and urbanism. MIT Press, Cambridge, MA
Collinson D (1985) Ethics and aesthetics are one. Br J Aesthet 25(3):266–272
Cross N, Dorst K, Roozenburgh N (1992) Research in design thinking. Delft University Press, Delft
Dean J (2002) Aesthetics and ethics: the state of the art. Aesthetics. Retrieved: http://aesthetics online.org/articles/index.php?articles_id=15
Dorst K, Cross N (2001) Creativity in the design process: co-evolution of the problem solution. Des Stud 22(2):425–437
Ellin N ([1996] 1999) Postmodern urbanism. Princeton Architectural Press, Princeton
Fainstein SS (2010) The just city. Cornell University Press, Ithaca
Fischler R (2012) Fifty theses on urban planning and urban planners. J Plann Edu Res 32(1):107–114
Fox W (2006) A theory of general ethics: human relationships, nature and the built environment. MIT Press, Cambridge, MA
Friedmann J (1982) Good society. MIT Press, Cambridge, MA
Hausman CE (1975) A discourse on novelty and creation. Martinus Nijhoff, The Hague
Hillier B, Musgrave J, O'Sullivan P (1972) Knowledge and design. In: Mitchell WJ (ed) Environmental design: research and practice EDRA 3. UCLA, Los Angeles
Hirt SA (2009) Premodern, modern, postmodern? Placing new urbanism into a historical perspective. J Plann History 8(3):248–273
Howard ZP (2002) New urbanism: a new approach to the way America builds. Spatium, pp 22–27. Retrieved: www.revistasusp.sibi.usp.br/scielo.php?pid=S0104-609
Le Corbusier (1923) Vers une architecture. Flammarion, Paris
Lindblom CE (1965) The intelligence of democracy: decision making through mutual adjustment. Free Press, New York
McKay S, Murray M, Macintyre S (2012) Justice as fairness in planning policy-making. Int Plann Stud 17(2):147–162
Newell A, Simon HA (1972) Human problem-solving. Prentice Hall, Englewood Cliffs
Rand A (1943) The fountainhead. Bobbs-Merrill, New York
Rondanini N (1981) Architecture and social change. Heresies II 3(31). Retrieved from http://en.wikipedia.org/wiki/Architecture#Historic_treatises
Schön DA (1983) The reflective practitioner: how professionals think in action. Basic Books, New York
Talen E, Ellis C (2002) Beyond relativism: reclaiming the search for good city form. J Plann Edu Res 22(1):36–49
Wikipedia (2012) Retrieved: http://en.wikipedia.org/wiki/Design
Williams RJ (2004) The anxious city: English urbanism in the late twentieth century. Routledge, Abingdon
Wolford JN (2004) Architectural contextualism in the twentieth century, with particular reference to the architects E. Fay Jones and John Carl Warnecke. Ph.D. in architecture dissertation, Georgia Institute of Technology. Retrieved: http://smartech.gatech.edu/handle/1853/7032
Wotton H (1624) Vitruvius: on architecture. Retrieved from http://en.wikipedia.org/wiki/Architecture#Historic_treatises
Wurster CB (1965) The social front of modern architecture in the 1930s: more than a "style"? Promising principles unfulfilled. J Soc Archit Historians 24(1):48–52

Chapter 2
A Conversation About Who's In? Who's Out? And Who Answers Those Questions When Planning for and Designing the Downtown

Carol D. Barrett

2.1 Introduction

> The tragedy of urban design today is that the effect on the quality of life of people of what is actually built is not given the emphasis that it deserves (Edmund Bacon, 1967)

Two different incidents in the author's life created the foundation for this conversation about the ethical responsibilities we have as design professionals when planning new spaces in a downtown. In the first instance, as a graduate planning student, the author participated on a team engaged by Atlanta (GA) parks staff to propose a redesign for a downtown park that would be unwelcoming to the homeless currently spending nights and days sleeping there. The solution involved nothing more elaborate than dirt and grass. If the site were graded into knolls, there would be places to sit and enjoy noontime park events. But these same undulating features would prove inhospitable to sleeping, especially after an evening dampening by the irrigation system. Visiting the park years later, the grassy berms are still in place. More traditional seating had been added as well, a good thing for those unable to easily sit down and get up from the grass, something the students had never considered.

In the second instance, while working in Berkeley (CA), the author came to know the names and faces of the homeless men who slept in the office entryway and used the city's restrooms and offices throughout the day. The competition for warm doorway space was intense. The homeless would appear with bedrolls and shopping carts a little after 5:00 pm to stake their claim. By 6:00 pm the doorway was filled with individuals settled in for the evening. When leaving left the building, the author would open the door just a few inches and call out a greeting. The homeless would shift their bags to

C.D. Barrett (✉)
Community Development Department, City of Burbank, CA, USA
e-mail: caroldbarrett@gmail.com

C. Basta and S. Moroni (eds.), *Ethics, Design and Planning of the Built Environment*, Urban and Landscape Perspectives 12, DOI 10.1007/978-94-007-5246-7_2,

allow the author's exiting; and friendly words were exchanged as the author left the building. The use of the restrooms by the homeless caused the greatest consternation. The bathrooms were used so thoroughly for toileting and bathing that some employees felt the bathrooms were unsanitary. The city increased the frequency of cleaning. That helped some. However, several of the more fastidious staff members called for sterner measures. A compromise was reached with the employees' union which raised a grievance. The first floor restrooms would remain accessible to the general public including the homeless. A combination lock was installed on the second and third floor restrooms. Those facilities retained their sanitary and tidy character.

During the intervening years between these events, the author had learned to think very differently about the homeless. The homeless have rights which must be accommodated as professionals help design places like the downtown. It's a matter of social justice. The homeless always had those rights; the author had failed to recognize them.

Not all design professionals will agree. Social justice is sometimes pushed to the back burner when planners and designers are charged with the task of planning for downtowns. Several years ago, a newspaper story was published about a large Texas city enjoying a thriving and growing downtown with new residential and office complexes. The reporter focused on restaurants emerging from vacant storefronts. The neighborhood was in transition, although it remained within easy walking distance of the city's premier center for homeless services and accommodations. The newspaper photograph documented the transition. Besides the new eateries, weedy open spaces had been replaced with flower gardens. The story of volunteer efforts and self-financed gardening of abandoned spaces was impressive. Toward the bottom of the story, it was briefly mentioned that the garden of riotous color previously had been a homeless encampment for seven people. Where had the seven people gone? The reporter hadn't thought to include that information in the story.

The downtown, where the homeless or underemployed often congregate, is not always conceived of as a battleground for social justice, but it can be if planners, landscape architects, and architects are not attentive to the fundamental matter of who gets to use the space. In a nation where private property rules predominate, the homeless (whose private property consists of a few meager items subject to police confiscation) have little to no private property and therefore have no protection. They are often threatened by efforts to improve the neglected or forgotten parts of the public realm—the only space that is theirs.

2.2 The Process

Planning for downtown development means helping to choose who will be regulated and what form those regulations will take, though that is rarely the primary motivation of communities. Our downtowns provide space for public life. While residents live, work, and shop in different neighborhoods, the downtown is a shared physical reality with the presumed shared right to use that space, irrespective of income.

Many downtown plans conceive of the public as homogeneous, compliant, and reasonable. The unspoken assumption is that the public has a shared interest in a downtown setting that is interesting, safe, and attractive. Most of the downtown plans also accept that the right to public space imposes on us the duty of appropriate public behaviors toward others in that space. Downtown plans are sometimes accompanied by "quality of life" initiatives that seek to regulate street behavior, sleeping in public, and panhandling. Through these laws and other means, cities seek to use a seemingly stable, ordered urban landscape as a positive inducement to continued investment and to maintain the viability of current investment in core areas. But initiatives that criminalize sleeping in public do not address root causes such as the lack of housing, structural unemployment, and the despair of addictions. The problem is much more profound than individual disorder.

Consider the list of those often "missing in action" when downtown development plans are formulated: street vendors (unless it's coffee or sushi that is being sold from a tastefully designed cart), street performers whose music and mime are presented for tips, protestors and other concerned citizens, teenagers who seek a hangout, picketers, panhandlers, and the homeless. Depending on the size and location of downtown, there may be hustlers, prostitutes, and drug dealers who also conduct business in the public right-of-way or its alleys and corners.

2.3 The Content

For purposes of constraining this conversation to a manageable length, assume a specific task: Prepare a design-oriented plan and construction drawings for Downtown A that focuses exclusively on the public realm. Downtown A has wide sidewalks, but no typical amenities. Alleys are used only for deliveries and dumpsters. The downtown courthouse square itself has not been updated since the 1950s when the courthouse itself was last renovated. Existing small businesses, bars, and eateries, however, are doing reasonably well due to the presence of daytime government employees and a vibrant bar scene in the evenings because of a nearby college.

The professional work begins with a series of questions: (i) What kinds of spaces are in the downtown? (ii) Who uses the space? (iii) Who does not? (iv) Who is welcome? (v) Who is not?

One can gather the data by observation and by asking both daytime and nighttime users and interviewing knowledgeable service providers and business owners. The police know where all of the downtown's varied users can be found: the culverts and overpasses where people sleep, the hangouts for teenagers or those with substance dependencies, or the corners where panhandlers most effectively work the crowds. Table 2.1 shows the outcome of the process of inquiry for Downtown A. With all of this data you are better able to answer the central question of how the needs of all current and potential users will be addressed in the process of planning for and allocating downtown spaces.

Table 2.1 below identifies the types of open and vacant spaces in Downtown A. The last column indicates how the spaces are used and the extent to which the use or users are welcome. This is an expression of the underlying values. In Downtown A the spaces are used mostly for specific activities rather than for spontaneous social

Table 2.1 The public-private space continuum in Downtown A

			How is the space used?
Type of space	Who owns?	Who maintains?	Who is welcome and why?
Streets	Public	Public	Everyone with a vehicle mobility requirement is allowed access to the streets—cars, bicycles, buses, etc. Occasionally access may be limited for special events. Permits can be granted for parades and protests. Provided there are sidewalks, those without vehicles will find their use limited to crossing the street. Walking outside the permitted boundaries (jaywalking) can earn a pedestrian an expensive citation. Similar prohibitions apply to those whose mode of transportation is not "street legal," such as skateboards and mini motorbikes
Parks	Public	Public or public improvement districts	Public, but expected to be orderly and reasonable. Used for mobility, casual meetings, civic life, entertainment, people watching, etc. Includes homeless and restless teens who may be only tolerated
			Used for festivals, events, or other less well-accepted activities like drinking by transients and day stops for homeless
Sidewalks and open space around public buildings	Public	Public in general. Sometimes food service businesses also are required to keep front area clean	Public, but expected to be orderly and reasonable. Used for mobility, casual meetings, civic life, entertainment, people watching, etc. Includes homeless and restless teens who may be only tolerated. Use may be impeded by newspaper machines, dining tables, tent signs, etc., all of which limit the public's use for mobility
		Sometimes licenses to use sidewalks for food service are granted and also carry requirements for maintenance	May entail closer regulation of acceptable behavior based on regulations being developed and enforced by the private sector

		Public improvement districts—special assessments of private property to maintain public areas to higher standards of cleanliness	
Vacant land/abandoned buildings	Primarily private but could be public due to tax foreclosures	Often not maintained	Not designed to be used but provide important spaces for hidden populations—homeless, transients, teens. Used for hanging out, camping, etc.
Private open space like plazas	Property owners	Property owners, but there is often a public agreement on design, use, and maintenance standards	Customers and the public,[a] provided the public's use does not impede use by customers. Used for resting, dining, and people watching

[a]The essential difference between a customer and the public is that customers are welcome because they have money to spend. The public is welcome irrespective of income

contact between strangers and dialogue among citizens as may have been the expectations of the architects who designed the courthouse square a hundred years before. Many downtown users have a simple requirement to be able to pass through—to be a stranger among strangers and to move safely to conduct business. Downtown A is a bit different from many because of the highly social nature of the student population and the potentially social nature of interactions at the courthouse bringing together a good cross section of the community. Together, all these groups create an active environment.

Conversely, some downtown users may make others in the community feel less comfortable. For example, the homeless congregate in specific areas causing some to question their own security and comfort. However, it is the presence of the kindhearted and generous students who make the downtown so attractive to panhandlers and the homeless. It is the issue of the perceived safety of the downtown that has prompted this plan as the worsening economy has increased the number of visible homeless and behaviors that are called problematic.

Another example of a group often targeted for displacement is youth. Teenagers have no obvious right to private spaces, so they congregate in public places. They choose spaces adults do not want, such as parking lots and other isolated areas. Their behavior, which constitutes loitering and rowdiness in the minds of some, also can affect one's comfort level. Skateboarding is perceived by others as noisy and actually can destroy public property like walls and benches as the more skilled boarders incorporate street furniture into their rides. Downtown business owners are anxious that the plan somehow also addresses the presence of the teenagers and their activities.

After establishing a clearer picture of how the downtown is used and by whom, design professionals should work to include all of the perspectives in the design phase. It would be best to have all of these groups participate directly in the planning, but this may be difficult to achieve. The designer has few incentives to offer, and there is much distrust to overcome. People struggling with difficult issues of survival may not find an evening workshop with cookies a compelling reason to participate. But professionals should try. With some assistance or information, some of the not traditionally welcomed users may be willing to engage if the venue for their engagement feels safe. Sometimes service providers must act as stand-ins.

After a full community engagement process, the professional is charged with the task of crafting the outcome. The programs, alternatives, and policies outlined in the plan should recognize the legitimacy of uses by all the residents, including panhandlers, teenagers, and the homeless. In creating downtown plans, some cities have funded resource centers for the homeless, built transitional housing, constructed skate parks for disaffected teenagers, and provided public spaces for street performers. As funding for services has declined, the competition for those resources has become more intense. The design professionals can assist in brokering compromises among users.

The policy questions to ask for each of the proposed plan outcomes include who benefits and who loses. A second and more value grounded set of questions then emerge: What outcomes do we want and to whom should the benefits be targeted? Who is at the decision-making table and will they have any power? This last question may turn the starting point for downtown planning on its head.

2.4 The Values: Ethical Considerations

To what extent does the right to access the downtown—including the right to hang out on streets and sidewalks—belong only to people with specific commercial activities to transact? Design professionals become involved in endeavors to renovate and rehabilitate downtown, to upgrade, and to sanitize in some cases. The goal may be to create something more competitive with the mall, if the emphasis becomes creating a delightful shopping experience. In the process, public space may become privatized and regulated.

The worldwide recession has increased the scope and needs of those who lack access to jobs, housing, and services and who spend their days and nights in parks and on sidewalks and the rights-of-way. Complaints are made to municipal governments about aggressive or antisocial behavior including (i) using loud, profane, or threatening language; (ii) aggressive panhandling; (iii) blocking use of sidewalks by lying in the right-of-way or storing bulky personal belongings; (iv) defacing parks and sidewalks with garbage and drug paraphernalia; or (v) urinating and defecating in public.

Much of the behavior triggering the complaints may come from only a small percentage of those seen living on the streets. Of the people exhibiting the most troublesome behavior, many suffer from mental disabilities, have alcohol or other drug additions, or have dual problems. Some may also be homeless.

Seen from the other side, the very presence of people exhibiting troublesome behaviors in public space seems threatening to the safety, pleasure, and economic desires of some urban residents. So how does one respond when challenged by someone else's assertion to a place in the downtown or a voice in determining the future of downtown? Can design professionals themselves admit that they themselves are sometimes uncomfortable? A strong source of guidance in determining a professional response to the question of meeting the requirements of the homeless is a Code of Ethics.

The Australian Council of Professions puts forth a definition of profession appropriate for this conversation (Brigham 2009, p. 2):

> A profession is a disciplined group of individuals who adhere to ethical standards and uphold themselves to, and are accepted by the public as possessing, special knowledge and skills in a widely recognized body of learning, deriving from research, education and training at a high level, and who are prepared to exercise this knowledge and these skills in the interests of others.

Licensed professions have a legal, moral, and ethical duty of providing their expertise in the public interest. Providing special knowledge even to people who cannot pay for it is part of what elevates a business into a profession. The argument can be made based on professional licensure, which is a sanctioned monopoly granted to professionals on the provision that they provide service in the public interest. Providing pro bono services offers access to this knowledge to those who cannot afford it is part of a profession's service to the greater public interest.

All of the design-oriented professions have Codes of Ethics which, in some form, address our responsibility to serve the public interest and be attentive to the issues of social justice. The statements in the codes make the case for why design professionals must, despite their personal comfort level, try to figure out a way to address the needs of the homeless as part of the downtown plan.

The 2007 *Code of Ethics and Professional Conduct* of the American Institute of Architects includes canons which are broad principles of conduct, ethical standards which the architects define as more specific goals toward which members should aspire, and the rules of conduct which are mandatory. Several elements of this code speak to the values and ethics issues at play in this conversation and are summarized below.

> *Canon 1.* Members should ... thoughtfully consider the social and environmental impact of their professional activities.
>
> > E.S.1.3 Member should...improve the environment and the quality of life with it.
> > E.S.1.4. Members should uphold human rights in all the professional endeavors.
> > E.S.2.2 Members should render public interest professional services including pro bono services, and encourage their employees to render such services. Pro bono services are those rendered without expecting compensation, including those rendered for indigent persons, after disasters, or in other emergencies.
>
> *Canon 2.* Members should promote and serve the public interest in their personal and professional activities.
> *Canon 6.* Members should promote sustainable design and development principles in their professional activities.

The *Code of Professional Ethics for the American Society of Landscape Architects* follows the same nomenclatures as the American Institute of Architects: canons, ethical standards, and rules. The component of the landscape architect's ethics that most closely relates to our conversation is

> E.S.2.1 Members should understand and endeavor to practice the objectives and strategies of the Declaration on the Environment and Development.

One of the unnumbered objectives is to:

> Actively engage in shaping decisions, attitudes, and values that support human health, environmental protection, landscape regeneration, and sustainable development.

If one reads the Declaration, additional guidance is provided:

> In developing landscape architectural design, planning, management, and policy projects, identify and involve stakeholders—both communities and individuals—in helping to make decisions which affect their life and future; ensure that they have appropriate access to relevant information, presented in an understandable form; create opportunities for them to contribute to solutions.
>
> Commit to solving problems within the site; don't transfer problems or postpone solutions.
>
> Develop and share information which helps define the issues or contributes to solutions that focus on sustainable and equitable development.

The *Code of Ethics* first adopted by American Institute of Certified Planners in March 2005 clearly lays out a set of expectations for certified planners:

1. Our Overall Responsibility to the Public (Excerpts)
   (a) We shall always be conscious of the rights of others.
   (f) We shall seek social justice by working to expand choice and opportunity for all persons, recognizing a special responsibility to plan for the needs of the disadvantaged and to promote racial and economic integration. We shall urge the alteration of policies, institutions, and decisions that oppose such needs.

"We shall seek social justice." Those five simple words in the AICP Code of Ethics can trigger deep emotional and intellectual responses. At its core, the ethical standard imposes a duty on planners to advocate for those whose voices are not being heard. A short list would include children, the poor, transients, refugees, the very ill, or the elderly. The code's exhortation to seek justice should be one that underlies organizing and carrying out a downtown plan. It is worthwhile to note that the specific language adopted by the AICP Commission uses an action verb: "seek." It does not say, "Give some limited consideration to social justice." The AICP Code of Ethics also reflects the expectation that planners will urge changes in policy to achieve a better economic and racial integration. The code for planners requires a direct and affirmative response. It sets a higher bar than the "consider social impact" of other codes.

The *National Society of Professional Engineer's Code of Ethics* also has persuasive components. The term equity appears in the Preamble as a basic requirement of the engineering service. Both the first fundamental canon and the first rule of practice call for the public health, safety, and welfare to be paramount. The outline of professional obligations includes

III.2. Engineers shall at all times strive to serve the public interest.
III.2.a. Engineers are encouraged to participate in civic affairs... and the well-being of their community.
III.2.d. Engineers are encouraged to adhere to the principles of sustainable development in order to protect the environment for future generations.

See Table 2.2 which compares the codes and their expectations for design professionals. The language is clear in terms of addressing the broader needs of a community which includes the homeless.

So if there is an expectation embraced by design professionals through their Codes of Ethics to consider issues of equity, why wouldn't there be a stronger expression of social justice toward the homeless in every plan for the downtown? Possible reasons include:

1. The design professional is not a member of the professional society and so is not bound by its Code.
2. The design professional doesn't find the provisions persuadable or enforceable.
3. There are other aspects of pro bono work which may require design professionals to operate outside their traditional comfort zone. This includes being open to more unconventional collaborative partners and using creative skills to solve process as well as design problems.

Table 2.2 Ethics/values tied to the treatment of the homeless that should emerge in planning for a downtown public realm based on existing ethical codes and principles

Ethic/value	Engineers	Architects	Planners	Landscape architects
Equity	Yes	Uphold human rights	Expand choice and opportunity	Contribute to solutions that focus on equitable development
		Consider social impact	Give people impacted by plans a chance to help prepare them	Accept responsibility for consequences of work on natural systems and cultural communities and their harmony, equity, and balance with one another
			Include those who lack influence	
			Seek social justice	
Pro bono service	Encouraged to work in civic affairs and work for the well-being of their community	Members should be involved in civic activities as citizens and professionals	Contribute time and effort to groups lacking in adequate planning resources and to voluntary professional activities	Encouraged to serve on elected or appointed boards, committees, or commissions dealing with the arts, environment, and land-use issues

Note: "Yes" means that Code language is explicit. Narrative lays out the language that most closely ties to the value

4. Pro bono service can also test many design professional's traditional skills. The ability to work creatively within a limited budget, an accurate knowledge of construction methods and associated costs, and due diligence to ensure projects are designed with longevity, adaptability, and low maintenance in mind are all much more important in pro bono settings.
5. There are additional challenges when planning and design services are offered through university-based studio projects:

- Difficulties programming community service work around academic timetables.
- Difficulties in enlisting professionals to provide oversight.
- Difficulty of meeting significant community-based demand while completing other subjects.
- Academic staff may not always have the appropriate skills to manage a project.
- Most universities are in an urban environment. The cost of living is too high to expect students to further extend their academic financial commitment to work on lengthy community service projects (Brigham 2009, p. 36).

6. The Code provides an "out" for the professional.

- The AICP Code of Ethics offers a safe harbor for avoiding issues of equity or social justice in its provision A.2.(b): We shall accept the decisions of our client or employer concerning the objectives and nature of the professional services we perform unless the course of action is illegal or plainly inconsistent with our obligation to serve the public interest.
- The engineer's code offers a bit more ambiguous guidance in I.4: Act for each employer or client as a faithful agent or trustee.
- The architects may choose to act under Rule 3.103: Members shall not materially alter the scope or objectives of a project without the client's consent.

7. Work overload and the rate of changing practice and technology may leave little time for deeper learning and reflection upon ethical responsibilities (Gardner 2007, p. 227).
8. Individuals chose to be attentive to some ethical responsibilities and assume that others will take up the remaining requirements.

Of the list above, the most persuasive reason for practicing professionals is that the public service values inherent in the professional codes may be in conflict with the values of downtown business owners and some elected officials. These values are often rooted in the idea that the purpose of the entire downtown public infrastructure is to bring customers to the front door of businesses and these customers will spend money and generate sales tax. A secondary value is that those customers should have pleasant enough time spending their time and money that they will wish to return. The public realm should help contribute to this pleasant experience. These values are grounded in the belief that the promotion of consumption is an appropriate role for government. Anyone who doubts the persuasive power of this belief should recall the exhortations of elected officials post 9–11 to go shopping in the aftermath of the disaster to prove the American peoples' patriotism. By implication, anyone or anything which negatively impacts the quality of that shopping experience is to be eliminated. People who are shoppers are desirable in the downtown. Others are not.

Those design professionals motivated by standards calling for serving the public interest or promoting human dignity do not assume that unequal access for some people is inevitable. Downtowns are built by people and their governments and can be made more responsive to the needs of all the people.

> While no one should minimize the potency of external pressures, in the end it is individuals who compose our institutions and professions. If they cannot or will not hold themselves accountable, then it is up to their peers to do so. Design professionals, after the standards of the professions are clearly articulated, could ostracize those who do not meet them. In the absence of such checks, the prognoses for the professions are bleak. (Gardner 2007, p. 265)

2.5 Testing the Ethics of Two Plans

Having fixed upon the idea that design professionals are required by their respective Codes of Ethics to plan for the needs of the homeless when undertaking plans for the downtown, two documents were examined for conformity to Code provisions.

2.5.1 Design Guidelines for Public Plazas

Typical guidelines for the design of public plazas may be tested against the ethical principles. A recently published urban design plan for a Los Angeles neighborhood was examined for this purpose.[1] The only provisions found which in any way relate to the codes include (i) all design elements shall meet initial cost, long-term maintenance, management requirements, and sustainability goals; (ii) provision of universal access and safety shall be fundamental considerations in material and design solutions; (iii) sustainable practices, methods, and materials should be employed in all aspects of the design of all public realm elements.

Design professionals in this case are already considering the larger context of equity as it applies to accessibility. But much more needs to be done. There are ways of organizing the work on an urban design plan to broaden the conversation that were discussed in Table 2.1 and further below.

2.5.2 A Downtown Plan

If any city can be relied upon to embrace responsibility for the needs of the homeless in the downtown, it would be Berkeley, California. After a 6-year effort and nearly 200 public hearings, their downtown plan was adopted in 2012 and includes a chapter on community health and services. The Berkeley plan was selected for analysis because it was while working in Berkeley that the author came to a full appreciation of her responsibilities as a planner bound by the tenets of the AICP Code of Ethics. Another reason for selecting the Berkeley plan is that Berkeley's downtown shares similar features with the original Downtown A: a major civic square, student populations with an active restaurant, and entertainment district. Does the Berkeley plan live up to the expectations and exhortations of the various codes? Does it answer the question about ethics informing the planning process and provide a model for Downtown A? "Yes" is the answer to all three questions. Some of the policies most pertinent to the discussion about meeting the needs of the homeless are listed below. Others appear in Appendix A:

> *Goal HC-4.2*: Affordable Housing and Supportive Services: Promote the creation of permanent affordable housing with supportive services in Downtown, especially for homeless individuals and families. Encourage provision of appropriate supportive services for tenants at all functional levels.
>
> (a) Identify opportunities to expand permanent housing with supportive services in Downtown.
> (b) Develop programs and partnerships among service providers and non-profit housing developers for rehabilitating and converting existing SRO* properties....
> (c) Identify sites and long-term funding to support the development and on-going provision of services for new permanent supportive housing to meet the needs of very low-income single individuals....
> (d) Encourage the creation of "micro-units," very small apartments that many not include typical apartment features, such as a standard kitchen. Review development standards... to identify obstacles to the creation of micro-units, and consider whether such obstacles should be removed.
>
> *Goal HC-5*: Deliver in downtown effective and compassionate services for seniors, parents and youth, and persons with special needs, including individuals who are homeless, have physical and/or mental disabilities, and/or suffer from substance abuse.
>
> *Goal HC-5.4*: Social Services. Maintain and enhance prompt access to social services by Downtown residents and transient populations.
>
> (a) Evaluate existing and future social service needs and opportunities, both citywide and Downtown. Consider how services might be improved and how they might be accommodated in Downtown.

The plan recognizes that providing deeply affordable housing is one way to reduce homelessness. A number of policies focus on this aspect.

> *Goal HC-3*: Offer diverse housing opportunities for persons of different ages and incomes, households of varying size, and persons of varying abilities.
>
> Give downtown a significant role in meeting Berkeley's continuing need for additional housing.
>
> The Berkeley response and commitment may not be achievable in every community. But there are options for improving every downtown planning effort that should be explored.

Table 2.1 laid out one way to approach the issue: Begin with an analysis of who is welcome in the downtown. The elements of the table are based on a place where the author worked. After creating a table of users for work as a first step, the next step requires one to consider a number of questions about how to involve all of the users, how the plan will be made, and what will be included in the plan.

2.6 The Process of Making the Ethical Plan

2.6.1 The Background Work

As discussed earlier, the simple and most effective tool for raising the ethical bar when planning and designing is asking questions.

1. Who is making decisions? Was the process inclusive and did it empower everyone?
2. What criteria are used to decide on the elements of the plan? Are they the right ones?
3. Will the implementation process improve transparency? Is the work of implementation shared and understood? Will the outcomes be publicly reported?

The somewhat more complicated part, it must be acknowledged, is answering the questions.

2.6.2 Deciding What Goes into the Plan

Much of what goes into a plan is derived from any standard scope of work: vision and goals; data collection and analysis; development and selection of alternatives; and implementation strategies all informed by citizen participation. Bumping up the ethics quotient involves a more thorough community engagement process and broadening topics considered. Social Economic Environmental Design Network (SEED) has published a list of topics that identify elements of ethical content. That list appears in Appendix B. An explanation of SEED appears later.

2.6.3 Understanding the Plan's Impact

1. Who benefits from the plan? Could others benefit?
2. How are the benefits distributed to people and places? Could they be distributed differently in a way that provides more equity?
3. Does the plan maintain the status quo? Is that a good thing? Should there be change?
4. Are rights and opportunities for all enhanced? If not, what changes are needed?
5. What are the consequences of the plan, either intended or unintended?

Answer the questions, and then consider what alternatives might reasonably be devised for addressing the needs of the homeless. See Table 2.3 below for some ideas.

Table 2.3 Actions/activities that may meet the needs of the homeless

What are key issues?	What can be done on-site?	For individuals somewhere in the city if not in the plan area
Safety/security	Convertible structures such as bench for limited shelter at night	Shelter
	Storage lockers included with other plaza amenities	
	Design wall treatments that can provide wind shelter at night	
Shelter	Camping areas	Camping areas
	Restrooms with showers	Restrooms with showers
	Relaxed enforcement of bans on sleeping in public places at night	Transitional housing with supportive services
Toilets and showers	Restrooms with showers	Restrooms with showers
		Restrooms open to the public in parking garages and other public buildings. Freestanding public toilets sited in the downtown
Food	Food delivered	Soup kitchen
	Edible landscaping	
Support services	Engagement in the planning process	Continuum of care including health, rehab, and employment
	Daytime resource center or resource workers present in the park	Community courts for alternative enforcement and adjudication

2.7 Innovations in Ethical Design and Planning Practice

Some organizations already embrace and act upon the professional ethics for the design professions. These innovative practices are models. The three main models for providing community service are university-supported design studios, pro bono services, and the traditional Community Design Centers. The American Institute of Architects recognizes multiple types of pro bono services (AIA 2001, *institute guidelines*): (i) offered by an individual, (ii) reimbursement for direct expenditures, (iii) reduced fee, (iv) assistance from AIA chapters, (v) offered by a firm, and (vi) offered by interns.

Organizations providing pro bono architectural and planning services include the following:

Community Design Centers are not-for-profit agencies and as such are able to make the provision of community service their core business. While they have great expertise in this field and unparalleled relevance, like other not-for-profits, they are dependent on funding from corporate benefactors, foundations, and government agencies. For CDCs to survive, they must have highly developed fund-raising capabilities and are subject to the vagaries of economic cycles. The American Institute of Architects, in 1977, established a national network now known as the Association for Community Design, and it continues to provide support to CDCs throughout the United States. Of the existing Community Design Centers active today, only a few have ties to those established in their heyday in the 1960s. There appears to be a pattern of centers forming in response to a particular event or around a strong individual and then closing or evolving into something else in response to external factors, mainly changes in personnel or funding.

Public Architecture is a not-for-profit organization with a principal mission to "put the resources of architecture in the service of the public interest." The organization has both promoted the cause and connected firms with those in need of design assistance. In 2005, the 1% Program was created by Public Architecture to challenge architecture and design firms to pledge 1% of their time to pro bono service.[2] As of June 2011, 1,048 architecture and design firms joined the program pledging more than 311,496 h of pro bono service.[3] In the United States, the main professional organization through which pro bono services are volunteered is Architecture for Humanity. Established in 1999 and based in San Francisco, it has a network of 40,000 professionals in 28 countries around the world. It relies on donations, grants, sponsorships, and volunteers and has completed more than 250 projects varying from orphanages to housing around the world. International projects are staffed by volunteer designers who live in the overseas location from concept to completion.

Community Planning Assistance Team (CPAT) is an initiative of the American Institute of Certified Planners. Multidisciplinary teams partner with community members to help nurture community empowerment. The program was established in 1995 by the American Planning Association.

Several organizations help to focus thinking about issues of equity and social justice.

Architects/Designers/Planners for Social Responsibility (ADPSR) has been working for social justice as one of its core missions, originally established in 1981 to promote nuclear disarmament; since 1990, ADPSR has focused much of its efforts on ecologically and socially responsible development.

Planners Network is an association of professionals, activists, academics, and students involved in physical, social, economic, and environmental planning. Both groups have conferences and publications. Many of their members are academics and their impact on students is more significant than on practicing planners. The predecessor of Planners Network was Planners for Equal Opportunity founded in 1964 at the American Institute of Planners Conference in Newark, NJ.

Social Economic Environmental Design (SEED) was founded in 2005 by architects, designers, and other interested parties at the Harvard Graduate School of Design. SEED provides tools—a network and certification—that guide design professionals toward community-based engagement in design practice. It is a principle-based group of individuals and organizations sharing ideas. SEED principles include advocating for those who have a limited voice in public life and promoting social equality through discourse.

2.8 Final Thoughts

When the author worked as part of a student team to work on a design for a downtown park, the work met the client's objectives, but gave no consideration to the needs of the homeless who were an important park user group. Students are among the least likely to raise issues of concern outside their direct charge. Students have a less well-formed notion of ethical obligations. It's not as if the assignment wasn't a worthy one: to create a space for downtown workers for noontime concerts. It was just an incomplete assignment. In another example of planning for the homeless, Berkeley, California, acquitted itself with high fidelity to ethical principles. (The author was not an employee of the city when the plan was prepared and deserves none of the credit.)

Why would the author, a practicing planner, write about this? It's the simple concept of the power of two. When there are matters of ethics and social justice that arise in the workplace, someone has to raise the issue. It is much more effective to have a conversation among like-minded people than a lecture from one thoroughly engaged but solitary individual. The more conversations, the more likely it becomes that colleagues will step forward to raise the questions of justice and equity.

This short conversation is grounded in values and questions about how outcomes can be changed when working on a downtown plan. What can be done to build downtowns that welcome and serve the needs of the homeless? Questions must be asked that redefine the downtown design issues and the possible solutions. It's not always easy, and sometimes it will not be possible to achieve a site-based solution. Action may have to be deferred to another place or time. But there is always something that can be done. In the short term, there are pro bono opportunities for service available to all design professionals. In the long term, the focus of change needs to be on the process by which we make the policies which frame the decisions.

All the design professions should adopt detailed and specific policy on pro bono and other community service activities. Code of Ethics should be amended to make specific the expectations for such service. The first step is an aspirational statement; specific conduct guidelines would be better. All design professionals should be willing to report the illegal or unethical conduct as required by their respective codes. A further step toward transformational practice would be for design professionals to note examples of less-than-ethical work, to derive lessons from those experiences, and to encourage others to pursue a more responsible course of action (Gardner 2007, p. 335).

Government and its consultants must pay attention to the needs of the poorest, most marginalized segments of society in the downtown. If the rights of the homeless to food and shelter are eroded, then increasingly the rights of everyone are at risk.

If the homeless are not visible, then people will never see the results of the society they are making. Design professionals should take a leadership role in recognizing and protecting these rights.

Notes

1. The urban design plan selected for review was submitted by the client and the firm which prepared the plan to the Los Angeles Section of the California Chapter of the American Planning Association. The plan was considered by some parties to be representative of best practices.
2. "If every architecture firm were to give one percent of their time, it would add up to 5,000,000 h which is the equivalent of 2,500 persons working full time for a year," according to the Public Architecture website.
3. Website for the 1% pro bono design program of Public Architecture. www.theonepercent.org/About/Participants.htm. Website viewed 6/5/12.

Appendix A Goals and Policies from the Berkeley Downtown Area Plan, 2012

Chapter 7: Housing and Community Health and Services, pp. HC-1 through HC-13.

Goal HC-3 Offer diverse housing opportunities for persons of different ages and incomes, households of varying size, and persons of varying abilities. Give downtown a significant role in meeting Berkeley's continuing need for additional housing.

Policy HC-3.2: Affordable Housing and Supportive Services. Encourage the creation of new affordable housing projects for low and very low income housing, and the creation of associated supportive services.

Policy HC-3.5: Senior and Disabled Housing. Encourage the creation of affordable housing for seniors and persons with disabilities, especially housing with supportive services, except for skilled nursing facilities that take little advantage of and contribute little to downtown's pedestrian and transit-oriented environment.

Policy HC-4.1: Prevent Displacement. Prevent displacement of existing affordable housing in the Downtown Area, except where replaced by an equivalent number of permanent similarly affordable dwelling units.

(a) Enforce and consider way to strengthen existing policies for the retention of existing rental housing for low-income residents.
(b) Maintain and enhance City "acquisition and rehabilitation" efforts for affordable housing, while avoiding arbitrary or capricious displacement of tenants. Mitigate the negative effects of temporary or permanent relocation on tenants, and develop a plan for such mitigations in advance of implementation.
(c) Consider incentives for the acquisition and rehabilitation of existing buildings by private owners, to maintain more affordable housing for low-income residents in the Downtown.
(d) Consider the use of the Housing Trust Fund and/or housing mitigation fees from office and other commercial projects for the renovation and retention of affordable housing.
(e) Consider flexibility in development standards to make it easier to renovate and retain affordable rental units.

HC-4.2: Affordable Housing and Supportive Services: Promote the creation of permanent affordable housing with supportive services in Downtown, especially for homeless individuals and families. Encourage provision of appropriate supportive services for tenants at all functional levels.

(b) Identify opportunities to expand permanent housing with supportive services in Downtown.
(f) Develop programs and partnerships among service providers and non-profit housing developers for rehabilitating and converting existing SRO* properties, and by using a permanent supportive housing model, such as Berkeley's "Housing First" program.
(g) Identify sites and long-term funding to support the development and on-going provision of services for new permanent supportive housing to meet the needs of very low-income single individuals and engage owners of SRO properties to convert to permanent supportive housing.
(h) Encourage the creation of "micro-units," very small apartments that many not include typical apartment features, such as a standard kitchen. Review development standards and inclusionary housing provisions to identify obstacles to the creation of micro-units, and consider whether such obstacles should be removed.
(i) Explore options for expanding the range of affordable housing opportunities in the Downtown by encouraging innovative housing types, including limited equity cooperatives, co-housing, housing land trusts, and other options.
(j) Consider incentives for projects that provide a greater number of affordable units or that provide units at deeper affordability (50% or less of the Area Median Income).

Goal HC-5: Deliver in downtown effective and compassionate services for seniors, parents and youth, and persons with special needs, including individuals who are homeless, have physical and/or mental disabilities, and/or suffer from substance abuse.

Goal HC-5.4: Social Services. Maintain and enhance prompt access to social services by Downtown residents and transient populations.

(b) Evaluate existing and future social service needs and opportunities, both citywide and Downtown. Consider how services might be improved and how they might be accommodated in Downtown.

Goal HC-6: Provide a safe, clean and attractive downtown, in partnership with the community.

(a) Establish community-appropriate standards of behavior and maintain a shared commitment among public and private stakeholders to enforce those standards, consistent with the city-wide Public Commons for Everyone initiative.
(b) Distribute public information summarizing existing ordinances pertaining to street behavior and provide clear instruction on how to report aggressive behavior, and unsanitary and unsafe conditions.
(c) Engage merchants, other stakeholders, the Police Department, mental health and social service providers, and homeless advocates, in defining critical issues and actions. As part of this on-going process, monitor locations and conditions where aggressive, abusive and unsanitary behavior occurs frequently.
(d) Provide adequate 24-h toilets in Downtown with clear signage, and provide for their ongoing maintenance, security, and frequent cleaning.
(e) Establish easy mechanisms for direct communication between Downtown community stakeholders and police or other service personnel to encourage rapid responses to unsafe conditions or inappropriate behavior.
(f) Work in partnership with Berkeley High School and its students, parents, teachers, and staff, along with merchants, to define what constitutes appropriate heavier—for students and adults alike—and to encourage appropriate behavior in Downtown.

Goal HC-7: Maintain and expand integrated health services available in downtown to address health inequities.

(a) Policy HC-7.1 Health Services. Encourage the retention and expansion of effective health care and health-related services in Downtown, especially to address the needs of those who would be most negatively affected by lack of (sic) accessible, centrally located health services.

SRO is the acronym for Single Room Occupancy.

Appendix B SEED Issue Identification

The SEED Issue Identification list below includes possible social, economic, environmental, and participatory issues addressed by a SEED project.

Social	*Economic*
Accessibility	Access to mainstream financing
Child care	Access to products
Civic engagement	Access to services
Crime and safety	Affordable housing
Cultural heritage	Business training
Education	Cooperative ownership
Elder care	Debt relief
Empowerment	Economic development
Equality	Economic education and training
Equity	Employment
Food security/hunger	Enterprise
Freedom	Entrepreneurship
Gathering spaces	Green collar jobs
Green gardening	Job security
Health	Job training
Housing/shelter—emergency	Living wages
Housing/shelter—homelessness	Micro lending
Housing/shelter—permanent	*Environmental*
Housing/shelter—transitional	Access to energy
Human rights	Access to nature
Learning	Alternative energy
Local identity	Biodiversity
Mobility	Conscious consumption
Organic gardening	Environmental education
Political activity	Environmental sustainability
Political planning and policy	Functional eco-systems
Prejudice/discrimination	Green energy
Rain water management	Local sourcing
Recreation/play	Environmental metrics: LEED, energy star, etc.
Strengthening community	Preservation of nature

(continued)

(continued)

Water	Preservation of wildlife
Water access	Public transportation
Water management	Sanitation
Well-being	Smart growth
Wellness	
Women/gender	
Participation	
Asset-based design	
Asset-based development	
Community charrettes	
Coordinated with local comprehensive plan	
Local government support	
Local media	
National government support	
Stakeholder advisory group	
State government support	

References

American Institute of Architects (2001) Institute guidelines to assist AIA members, firms and components in undertaking pro bono service activities, American Institute of Architects website, date of publishing 17 February 2009. Viewed 2 June 2011. http://www.aia.org/groups/aia/documents/pdf/aiab082967.pdf

Bacon EN (1967) Design of cities. Penguin, New York

Brigham C (2009) Enabling good deeds in design: how can architects provide community service as part of their practice. http://architectureinsights.com.au/media/uploads/resources/Callantha_Brigham_Report.pdf. Viewed 5 June 2012

City of Berkeley (2012) Downtown area plan

Gardner H (2007) Responsibility at work: how leading professionals act (or don't act) responsibly. Wiley, New York

http://www.publicarchitecture.org/The_1.htm

http://www.seed-network.org/evaluator/issues.php

http://www.theonepercent.org/

Website for Architects, Designers, and Planners for Social Responsibility. www.adpsr.org

Website for Planners' Network. www.plannersnetwork.org

Website for the American Institute of Architects. www.aia.org

Website for the American Institute of Landscape Architects. www.asla.org

Website for the American Planning Association. www.planning.org

Website for the National Society of Professional Engineers. www.nspe.org

Chapter 3
The Relevance of Public Space: Rethinking Its Material and Political Aspects

Stefano Moroni and Francesco Chiodelli

3.1 Introduction: Towards a Critique of the Two Mainstream Theses on Public Space[1]

The debate about public space is amply covered by a broad spectrum of disciplines: sociology, anthropology, architecture and planning and political sciences. Within this scenario, certain invariable focal points can be pinpointed that are common to a great deal of academic works and of journalistic and public debate also. The two recurrent theses on public space to which we refer are as follows:

1. Public space has a central role in the public sphere creation: "[public spaces] are spaces within which the 'public sphere' is formed, policed and contested" (Blomley 2001, p. 3); "[public space] provides a material basis for the public sphere" (Mitchell 2005, p. 85). "Theories of the public sphere … must always be linked to theories of public space. … The regulation of public space necessarily regulates the nature of public debate" (Mitchell 2003, p. 182).
2. Public space is subjected to a privatisation process: "It is practically a truism to say that the disappearance of public space is caused by privatization" (Kohn 2004, p. 4). This process is usually attributed to the development of new types of private settlements, for instance, contractual communities (like homeowners associations) and shopping malls.

The first thesis is generally employed to advocate the importance of public space. In this perspective, public urban space is considered important mainly for its own political value.[2] Adopting this view, it is asserted that the quality of a city is above all related to its public space: "the nature of public space … defines the nature of citizenship" (Mitchell 2005, p. 85). A corollary of this view is the idea that it is a

S. Moroni (✉) • F. Chiodelli
Department of Architecture and Urban Studies, Milan Politecnico, Milan, Italy
e-mail: stefano.moroni@polimi.it; smoroni@aznet.it; f.chiodelli@hotmail.it

C. Basta and S. Moroni (eds.), *Ethics, Design and Planning of the Built Environment*, Urban and Landscape Perspectives 12, DOI 10.1007/978-94-007-5246-7_3,

moral obligation, for example, for planners, to commit themselves in a battle to defend public space. This becomes even more pressing in the presence of a presumed progressive privatisation of public space (as the second thesis asserts): "Our commitment to free speech requires us to reconsider the spatial practices that can either enhance or inhibit that freedom. Most important, a proper understanding of the connection between spatial practices and freedom of speech should alert us to the dangers entailed by the erosion of public space" (Kohn 2004, p. 4).

In our opinion, however, these two theses are for several reasons quite inexact. We will present and defend two different theses that are in a certain sense opposed to the previous ones:

1. There is no necessary causal relation between public space and the public sphere; this is even more true nowadays – thanks, for example, to the development of the new information technology.
2. No privatisation process of public space is actually under way; on the contrary, an increasing collectivisation process of private space is in action.

It is important to highlight that our arguments also rest on the assumption that public space is *fundamental*. And, in fact, one of the purposes of this article is just that of confirming and strengthening this importance. In our opinion, this cannot however occur by using, as usually done, arguments that are not very persuasive, as happens with the aforementioned theses on the public sphere and of the privatisation of public space – they are a disservice to the cause. According to us, public space is not only important but also *necessary*. Nevertheless, this necessity is linked more to its "livability relevance" than to its "political relevance" – without any connection with the transformation processes that involve private urban spaces. The terms "livability relevance" and "political relevance" are employed here simply as labels to distinguish something that is relevant prevalently for certain physical, tangible actions (lingering in a square, sitting on a bench, moving from one neighbourhood to another…) from something that is relevant prevalently for certain "immaterial" aspects (communicating political messages, exchanging civic ideas…).

3.2 On the Concept of Public Space

The current discussion about public space is affected by a certain "Manichaeism". Quite often, public space is described as a space concerning the people as a whole (Ercan 2010). In this sense, public space is regarded as totally opposite to and different from private space. The latter is often considered like a space where it is possible to exercise an unlimited right of exclusion.

Actually, the situation is more complex. In Western cities, different articulations – that is, different ownership regimes – of both public and private property are present; in both cases, we have no necessary correspondence between *property* and *use*. There are private property spaces (e.g. bars) that can have more "collective use" (i.e. a use open to people, with few access and behaviour restrictions) than certain public property spaces (e.g. police stations).[3]

Strictly speaking, a public space can be distinguished from a private one on the basis of its owner: public space is a place where the owner is the State (the central State or the local governments), while private space is a place where the owner is represented by private legal persons. In this sense, public and private spaces are two clearly different and separate realities; nevertheless, this does not express per se what is the "publicness/collectiveness" of the space at issue. This specification is important because, as we will see in the following sections, it is possible to maintain – for instance – that the diffusion of some private settlement models (e.g. shopping malls and contractual communities) entails paradoxically an increment of spaces used in common, or that many public functions (connected with public sphere or political mobilisation) can take place also in private property space as well – and not only in the public streets or squares – without any necessary decline in "publicness".

In order to go on with our discussion, at any event, it is useful to break the categories of public spaces and private spaces (as spaces owned by different subjects) down into a set of subcategories. In particular, we can distinguish among six kinds of urban spaces in the following manner (Moroni and Chiodelli 2013): first, *stricto sensu public spaces* (i.e. public spaces of the connective and open type for general use: public squares and plazas, streets); second, *special public spaces* (i.e. public spaces assigned to special functions: public schools, hospitals, libraries, etc.); third, *privately run specific public spaces* (i.e. publicly owned spaces that are leased to a private subject: marinas, lidos, etc.); fourth, *simple private spaces* (i.e. private spaces for individual use: detached houses, etc.); fifth, *complex private spaces* (i.e. private spaces in which use is conceded only to a specific group of people, usually an association or club); and sixth, *privately owned collective spaces* (i.e. bars, restaurants, hotels, shopping centres, cinemas).

3.3 Two Different Theses on Public Space

3.3.1 First Thesis: The Non-necessary Overlap Between Public Space and the Public Sphere

As we have seen, a large quota of literature on the city focuses on the overlap between *public space* and the *public sphere*, that is, on the fact that public space is primarily important inasmuch as it is the place in which the public sphere develops.

The public sphere is usually defined, according to Habermas' well-known definition (1974, p. 49), as "a realm of our social life in which something approaching public opinion can be formed". Its nature is primarily "abstract", without a direct, necessary connection to (public) space: "it designates a theatre in modern societies in which political participation is enacted through the medium of talk" (Fraser 1990, p. 57). Habermas' public sphere idea is a-spatial.[4] The presumed necessary relationship between public space and the public sphere is stressed by (many) other sociologists, anthropologists and planners.

But if we define *public space* as a space owned by central or local state (i.e. as a space in which the rules of access and behaviour are determined by the public body) and the *public sphere* as an arena of public participation and deliberation, the two aspects/elements can in some cases overlap, but in other cases, this does not necessarily happen. The public sphere can in fact also develop outside public spaces. In the past, public space could have had a central role in the development of the public sphere, but, today, new technologies (e.g. the Internet) have weakened this role.[5] And this may be not necessarily negative: for instance, through new technologies, it is also possible to create "a dense web of sociality sustaining a civil society with a density and plurality of aims and objectives.... The net recreates [the] possibility of non-hierarchical discussion and free association" (Crang 2000, p. 309). In this sense, we can state that, nowadays, the public sphere is not *univocally* linked to public space.

This does not mean that a connection between public space and the public sphere cannot exist. Some connection between public space and the public sphere surely existed (Harvey 2006) and partially still exists (Lofland 2000). But there is nothing necessary about this, least of all necessary and sufficient. Nowadays, it seems that this connection is among varied types of space and the public sphere and not only between public space and the public sphere.

The physical places where public opinion nowadays takes shape are not only *stricto sensu* public spaces (e.g. squares and streets) or special public places intended to have particular functions (e.g. schools).[6] Dialogue and debate take place also in privately owned collective spaces: "Increasingly public life is flourishing in private places, not just in corporate theme parks, but also in small businesses such as coffee shops, bookstores, and others such ... places" (Banerjee 2001, pp. 19–20).[7] A considerable quota of contemporary urban society spends a lot of its time in private spaces of this kind, simply carrying out regular activities of interaction, socialisation and dialogue which in the past took place mainly in streets and squares.

Actually, political activities traditionally rely on the use of private space. Those "subaltern counterpublics" (Fraser 1990) that constitute the multiplicity of publics characterising contemporary societies often have their spatial location in private spaces. Habermas (1974) states that only "organized individuals" could take part and effectively participate in the process of "public communication"; these individuals can be organised in parties, associations, clubs, temporary groups, etc. Sometimes they use and occupy public spaces to demonstrate and to have successful political mobilisation; practically always they use and occupy private spaces for their everyday activities. As Kirby (2008, p. 83) argues referring to civil rights and non-violence movements, "important challenges to the status quo need not be restricted to the streets, and ... there has long existed an important tradition of political action occurring within privately-owned public spaces [e.g., bars]".

In the end, an important clarification is needed. It is important to warn against the dangers of nostalgia for certain public spaces of the past – spaces that were often far from ideals of inclusion, openness and publicness. Actually, spaces now viewed as embodying the ideal of democracy – for instance the Greek Agora – were usually also spaces of strong and violent exclusion (Dixon et al. 2006).[8] As Madanipour

(2010, p. 7) argues, there is "false romanticization of historic public spaces" on which we project our own political and social expectations.

3.3.2 Second Thesis: No Privatisation of Public Space

A great deal of critical analysis about contemporary urban development asserts that a privatisation process of public space is actually in action – even an "end of public space" (Mitchell 1995; Sorkin 1992; Low 2006). This "narrative of loss", which emphasises an overall decline of the public space (Banerjee 2001), is associated with the decline in the civic spirit and in social cohesion. Among the factors cited as principally responsible for all this are new forms of private spaces.

Contractual communities are considered one of the main perpetrators in this regard: "Gated communities represent a major reordering in the physical, social, legal and civic arrangements… The conversion of public to private space, inherent in gated community development, drives the process" (Lang and Danielsen 1997, p. 868). See also Blakely and Snyder (1997, p. 2): Today's homeowners associations "are not multi-unit, high-density apartment and condominium buildings with security systems or door-men in which gates or guards prevent public access to lobbies, hallways, and parking lots. Gated communities are different: their walls and fences preclude public access to streets, sidewalks, parks, beaches, rivers, trails, playgrounds – all resources that without gates or walls would be open and shared by all the citizens of a locality". Compare with Scott (1994, p. 20): "The assignment to homes associations of open space, parks, and other important community facilities bypasses the local governments that could appropriately be designated as custodians of such property".

This charge seems quite mistaken, however.

Let us consider, for example, the running of a homeowners association. This kind of contractual community is a residential complex whose inhabitants are members of an association: each member is the owner of his/her housing unit, and all members are co-owners of the common areas (streets, squares, parking lots, recreational areas, etc.). The members of the association accept preset rules on land use and pay an annual fee that is employed for managing the common spaces (Foldvary 1994; Nelson 2005). What happens in the creation of a homeowners association then is simply that a space privately owned (e.g. by a developer) is subdivided into spaces that are still private, some of which will be open to all the members of the future association. Hence, not only do homeowners associations not subtract any (previously) public space, but actually they organise (formerly) private spaces in a less parcelled method than the traditional way, encouraging the members of a certain group to use more of the common spaces: rather than the *privatisation of public space*, what happens here is a form of *collectivisation of certain private spaces* (Brunetta and Moroni 2012). In brief, and paradoxically, the phenomenon of homeowners associations "is causing an unprecedented transition from the traditional individual ownership of property to collective governance of most property in the USA" (Ben-Joseph 2004, p. 132).[9]

Contractual communities can clearly be criticised for several reasons. But the "public space privatisation argument" cannot be among them, because it is simply misleading.

Private commercial areas (e.g. shopping malls and outlets) likewise come under attack from the "privatisation" argument. In this case too, nevertheless, it is quite simple to observe how these structures do not entail any privatisation of spaces that were publicly owned before. On the contrary, they give citizens new spaces for collective use.

In the end, it seems possible to assert that many discussions about "public space privatisation" implicate a certain confusion. The risk is to confuse the (non-existent) reassignment of spaces that were publicly owned into private ones with the social and cultural transformation process which actually affects *the whole space*. The second process is really important – but it has nothing to do with any kind of "privatisation of public spaces".[10]

3.4 The Indispensability of Public Space

Public space is fundamental; *stricto sensu* public spaces, in particular, are indispensable to urban life. But the public space indispensability does not rest primarily on its "political" meaning. Surely, *stricto sensu* public spaces are the place where certain political interactions occur (Mitchell 2003) and where some political movements are visible (Mitchell 1995; Blomley 2001). But, in continually underlining this "political" meaning, sometimes we forget that space – and particularly *stricto sensu* public space – has also a fundamental "livability relevance" (Sects. 3.4.1 and 3.4.2).

It is important to note that stressing the "livability relevance" of public space does not imply assuming a reductionist position, that is, a position which devalues the immaterial meanings of public space. As we will see in what follows, however, certain symbolic-political meanings are in common with other types of space – not only physical ones. The political interaction, for instance, can occur in private spaces (e.g. bars) or in virtual locations too (e.g. social networks). On the contrary, the specific functional meanings we will talk about in the present section find their own expression *only* in *stricto sensu* public spaces. So it seems possible to argue that *stricto sensu* public spaces are *important* from an immaterial sociopolitical viewpoint, but *necessary* (indispensable) only from a "livability" perspective. And this "livability relevance" is – as we will see – a strictly *ethical* one.

3.4.1 Indispensability for People in a "No-Property Situation"

The mere fact that as individuals we have a physical body implies that to exist we must be in some place at any given time, and an absence of *stricto sensu* public spaces would entail that all those without private property could not "exist" or do anything. For certain categories of people in a "no-property" situation – such as the homeless – *stricto sensu* public spaces are what permit them to exist. As Jeremy Waldron (1993, p. 313) observes: "One way of describing the plight of a homeless individual might be to say that there is no place governed by private rule where he is allowed to be". The point is that "anything a person does has to be done somewhere. All actions involve a spatial component … It follows, strikingly, that a person who is not free to be in any place is not free to do anything; such a person is comprehensively unfree" (Waldron 1993, p. 316).

For those who do not possess a place of their own, *stricto sensu* public spaces are the only places in which they can carry out the functions of survival (sleeping, eating). For such a person, the prohibition of certain behaviours in public spaces – such as eating, sleeping or urinating – actually prevents him from these functions and as such prohibits him from "existing". The growing exclusion of certain forms of behaviour in public places[11] – without offering alternatives for carrying out such functions (public latrines, for instance) – means that people without private property, or without access to private services, are completely unable to carry out certain physical functions.[12] To quote Waldron (1993, p. 328) again: "If an action X is prohibited to everyone in public places and if a person A has no access to a private place in which to perform it, then action X is effectively prohibited to A *everywhere*, and so A is comprehensively unfree to do X".

3.4.2 Indispensability for All

The "livability" aspect of *stricto sensu* public spaces is fundamental also to people who privately hold – as owner or renter – a portion of urban space, that is, the great majority of people living in a city. Many examples can be considered here, but, for the sake of simplicity, we will focus only on one of them: the "connecting" function.

The "connecting" function of *stricto sensu* public spaces is really indispensable. Individual freedom is essentially based on freedom to move, that is, the possibility to move inside space: mobility is, in some respects, "constitutive of democracy"; it is a "democratic right" (Sheller and Urry 2000, p. 741).[13] Private property gives its owner a variety of powers and rights, but, paradoxically, it does not automatically entail the right to free mobility. You may own your own house, but if all around it there is private land on which you are not allowed to trespass, you are as if in jail, even if you possess your own jail. Even a libertarian like Robert Nozick (1974, p. 55) recognises this as a problem in a theory that gives absolute priority to private property rights: "The possibility of surrounding an individual presents a difficulty for a libertarian theory that contemplates private ownership of all road and streets, with no public ways of access. A person might trap another by purchasing the land around him, leaving no way to leave without trespass".

In the end, certain *stricto sensu* public spaces guarantee everybody the right to move from one point of the city to another, to reach other (public or private) spaces, where wished functions take place. In brief: "Public space mediates between the private spaces that make up the bulk of the city.... Without it, the spatial movement across the city becomes limited and subject to obstacles in need of constant negotiation" (Madanipour 2003, p. 220).

3.5 Conclusions

In this chapter, we have tried to argue two theses.

First, that public space is not perforce connected with the public sphere at all. As Kirby (2008, p. 91) writes: "There is nothing in our urban experience that demands that public space and the public sphere are inherently, ubiquitously and infinitely connected". Today, a large amount of daily interaction, meeting and communication no longer takes place only in public spaces. There are virtual or private spaces for collective use beyond these spaces. If we are interested in rebuilding the public sphere (and it is anyway doubtful that this has to be the planners' and architects' *central* purpose), action on (public) space does not appear as the better way nor the only one possible at all.

Second, we have argued that the privatisation process of public space is not actually under way. The opposite is happening: actually a "collectivisation" of certain private spaces is in place.

At the basis of our discussion, at any event, we retain the conviction that public space is an essential component of contemporary cities. The city could not exist without certain forms of public spaces, and these cannot be replaced by any suitable private space whatever. Necessary public spaces are first of all *stricto sensu* public spaces. Their own indispensability is, however, based primarily on questions of "livability". This does not mean that the "political" aspects of public space are irrelevant; it means merely that they must not be overplayed to the detriment of other fundamental roles played by public space. Clearly, public spaces that are made available and accessible for purposes of "livability" can turn out to be useful anyway even for "political" reasons (though the latter cannot be planned nor are they predictable or directly governable).

Notes

1. This article is the result of joint research activity undertaken by the two authors. The final written version of Sects. 3.1 and 3.3 can be attributed to Stefano Moroni and that of Sects. 3.2 and 3.4 to Francesco Chiodelli.
2. "Because by definition a public space is a place accessible to anyone, where anyone can participate and witness, in entering the public space one always risks encounter with those who are different, those who identify with different groups and have different opinions or different forms of life. ... Politics, the critical activity of raising issues and deciding how institutional and social relations should be organized, crucially depends on the existence of spaces and forums to

which everyone has access" (Young 1990, p. 241). "Publicly accessible spaces are important features of any vibrant and sustainable urban environment. The best spaces present opportunities for discussion, deliberation and unprogrammed, spontaneous encounters with those maintaining diverse viewpoint on the world" (Németh 2009, p. 2463).

3. To quote Low and Smith (2006, p. 3), public space "is not a homogeneous arena: the dimensions and the extent of its publicness are highly differentiated from instance to instance". In the same way, a private property space is not necessarily used by few persons – and in any case private ownership is never absolute, for it always includes duties and obligations (Needham 2006).
4. On this point, see Howell (1993, p. 311) and Mitchell (1995, p. 16).
5. As Sisk (2007, p. 1198) observes: "While the town square evolved in an era in which the primary means of communication was oral and most interaction was face-to-face, the opportunities for expression of ideas have expanded in number – and changed in nature – tremendously in the past several decades. The development of inexpensive access to a broad audience through internet technology promises to further revolutionize and democratize wide-ranging public debate in the future".
6. As Amin (2008, p. 6) argues: "Today ... the sites of civic and political formation are plural and distributed. ... Urban public space has become one component, arguably of secondary importance, in a variegated field of civic and political formation". See also Amin (2008, p. 5): "In the age of urban sprawl, multiple usage of public space and proliferation of the sites of political and cultural expression, it seems odd to expect public spaces to fulfil their traditional role as spaces of civic inculcation and political participation. We are far removed from the times when a city's central public spaces were a prime cultural and political site".
7. According to Light (1999), the origin of the public sphere in seventeenth and eighteenth centuries can be linked to private property space (e.g. coffee houses and salons). "It is possible to make the case that the public sphere has always been... part of an ongoing and ramifying development of congeries of semi-private social spaces" (Crang 2000, p. 309).
8. As Miles (2000, p. 255) notes: "In Attica in the time of Perikles, only twenty to thirty thousand people were citizens, all men, of a population of perhaps two hundred and fifty thousand; ... citizens alone participated in a democracy from which women, slaves and strangers were excluded". See also Basson (2006).
9. In other words: "At the end of the twentieth century, there was a ... shift in the United States from individual private ownership of residential property to new collective forms" (Nelson 2005, p. 351). See also Glasze et al. (2006, p. 2): "The value of 'public space' and its endangerment through 'privatisation' is a frequently cited *topos* within the critique of contemporary urbanism. ... [But] many master-planned private settlements simply involve the subdivision of a piece of land formerly under single private ownership into many titles under shared ownership. ... A piece of land under single private ownership may become co-owned by many residents".
10. Another point is worth clarifying. The fact that many collective activities take place in private spaces does not render them the opposite of "publicness". As Tyndall (2010, p. 134) writes, "too often urban research has framed publicity as a zero-sum game which, given the privatization occurring in our cities, is necessarily equated with a decline in publicness itself. ... Publicness is a social practice that is applied across a variety of spaces ... [and] is both constituted by, but also constitutive of space".
11. On this point, see for instance Mitchell (2003) and Laurenson and Collins (2007).
12. This does not mean that to allow a homeless individual to sleep on a bench is a desirable solution. This means simply that, in the absence of some form of public aid, to impede a homeless person from sleeping in a public space is *to prevent him from sleeping at all*. On this point, see also Mitchell (1997). As he observes, the "annihilation of (public) space" through a lot of restrictions as regards its use is a form of "annihilation of people". In Mitchell's opinion, anti-homeless legislation is not about crime prevention (as sometimes held) but about "crime invention".
13. For instance, in the USA, "freedom of travel can be invoked either as an implicit constitutional right or as a fundamental interest that triggers strict scrutiny under the Equal Protection Clause" (Ellickson 1996, p. 1239).

References

Amin A (2008) Collective culture and urban public space. Cities 12(1):5–24
Banerjee T (2001) The future of public space: beyond invented streets and reinvented places. J Am Plann Assoc 67(1):9–24
Basson S (2006) "Oh comrade, what times those were!" History, capital punishment and the urban public square. Urban Stud 43(7):1147–1158
Ben-Joseph E (2004) Double standards, single goal: private communities and design innovation. J Urban Des 9(2):131–151
Blakely EJ, Snyder MG (1997) Fortress America. Gated communities in the United States. Brookings Institution Press, Washington, DC
Blomley N (2001) Introduction. In: Blomley N, Delaney D, Ford RT (eds) The legal geographies reader. Blackwell, Oxford, pp 3–5
Brunetta G, Moroni S (2012) Contractual communities in the self-organizing city. Springer, Berlin
Crang M (2000) Public space, urban space and electronic space: would the real city please stand up? Urban Stud 37(2):301–317
Dixon J, Levine M, McAuley R (2006) Locating impropriety: street drinking, moral order and the ideological dilemma of public space. Political Psychol 27(2):187–206
Ellickson RC (1996) Controlling chronic misconduct in city spaces; of panhandlers, skid rows, and public-space zoning. Yale J Law 105:1165–1248
Ercan MA (2010) Less public than before? Public space improvement in Newcastle city centre. In: Madanipour A (ed) Whose public space? International case studies in urban design and development. Routledge, London, pp 21–50
Foldvary F (1994) Public goods and private communities. Edward Elgar, Aldershot
Fraser N (1990) Rethinking the public sphere: a contribution to the critique of actually existing democracy. Social Text 25–26:56–80
Glasze G, Webster C, Frantz K (2006) Introduction. In: Glasze G, Webster C, Frantz K (eds) Private cities. Global and local perspectives. Routledge, London, pp 1–8
Habermas J (1974) The Public sphere: an encyclopaedia article. New German Critique 3:49–55
Harvey D (2006) The political economy of public space. In: Low S, Smith N (eds) The politics of public space. Routledge, London, pp 17–33
Howell P (1993) Public space and the public sphere: political theory and the historical geography of modernity. Environ Plann D Soc Space 11(3):303–322
Kirby A (2008) The production of private space and its implications for urban social relations. Political Geogr 27(1):74–95
Kohn M (2004) Brave new neighbourhoods. The privatization of public space. Routledge, London
Lang RE, Danielsen KA (1997) Gated communities in America: walling out the world? Hous Policy Debate 8(4):867–877
Laurenson P, Collins D (2007) Beyond punitive regulation? Antipode 39(4):649–667
Light J (1999) From city space to cyberspace. In: Crang M, Crang P, May J (eds) Virtual geographies. Routledge, London, pp 109–130
Lofland L (2000) Urbanity, tolerance and public space. In: Deben L, Heinemeyer W, van der Vaart D (eds) Understanding Amsterdam: economic vitality, city life and urban form. Het Spinhuis, Amsterdam, pp 143–160
Low SM (2006) The erosion of public space and the public realm: paranoia, surveillance and the privatization in New York City. City & Society 18(1):43–49
Low S, Smith N (2006) Introduction: the imperative of public space. In: Low S, Smith N (eds) The politics of public space. Routledge, London, pp 1–16
Madanipour A (2003) Public and private spaces of the city. Routledge, London
Madanipour A (2010) Introduction. In: Madanipour A (ed) Whose public space? International case studies in urban design and development. Routledge, London, pp 1–15

Miles M (2000) After the public realm: spaces of representation, transition and plurality. J Art Des Edu 19(3):253–261
Mitchell D (1995) The end of public spaces? People's park, definitions of the public, and democracy. Ann Assoc Am Geogr 85:108–133
Mitchell D (1997) The annihilation of space by law: the root and implications of anti-homeless laws in the United States. Antipode 29(3):303–335
Mitchell D (2003) The right to the city. Social justice and the fight for public space. The Guilford Press, New York
Mitchell D (2005) The S.U.V. model of citizenship: floating bubbles, buffer zones, and the rise of the "purely atomic" individual. Political Geogr 24(1):77–100
Moroni S, Chiodelli F (2013) Typologies of spaces and topology of tolerance. J Urban Aff (forthcoming)
Needham B (2006) Planning, law and economics. Routledge, London
Nelson RH (2005) Private neighborhoods. Urban Institute Press, Washington
Németh J (2009) Defining a public: the management of privately owned public space. Urban Stud 46(11):2463–2490
Nozick R (1974) Anarchy, state, and utopia. Basic Books, New York
Scott S (1994) The homes association: will "private government" serve the public interest? In: Barton SE, Silverman CJ (eds) Common interest communities. Institute of Governmental Studies Press, Berkeley, pp 19–29
Sheller M, Urry J (2000) The city and the car. Int J Urban Reg Res 24(4):737–757
Sisk GC (2007) Uprooting the Pruneyard. Rutgers Law J 38(4):1145–1214
Sorkin M (ed) (1992) Variations on a theme park: the new American cities and the end of public space. Hill and Wang, New York
Tyndall A (2010) "It's a public, I reckon": publicness and suburban shopping mall in Sydney's southwest. Geogr Res 48(2):123–136
Waldron J (1993) Liberal rights. Cambridge University Press, Cambridge
Young I (1990) Justice and the politics of difference. Princeton University Press, Princeton

Chapter 4
Architects on Value: Reducing Ethics to Aesthetics?

Stefan Koller

4.1 Introduction

Historically, philosophical inquiries into ethics have repeatedly aligned with inquiries into architecture and, especially, the human city. Be it Plato's writings on the *polis* or Augustine's reflections on the two cities in his *Civitate Dei*, philosophical attention to matters at the heart of architectural practice seem not lacking (Illies and Ray 2009, p. 1199n.1). It may be surprising, then, that contemporary philosophy lacks its own sustained attempt at an ethics of architecture. With few notable exceptions (such as Scruton 1995), architecture does not receive the contemporary ethical attention it deserves. We may inquire why this is so, but also how to remedy this. One way of remedy would be to work one's way towards an ethics of architecture by starting from a layman's perspective on architecture. Another mode of remedy would seek to scrutinize extant ethical debates within architectural theory and see what can be done to improve these debates from a philosophical point of view.

The current work lies firmly in the second camp. This chapter scrutinizes a particular debate that is taking place within architectural discourse, on the topic of alleged reduction of ethics to aesthetics in architectural theory, and probes what a sustained philosophical reflection can offer to clarify and advance that debate. To do that, the chapter proceeds as follows.

In Sect. 4.2, the work offers for consideration sources which document and report the alleged reduction. It then attempts to formalize the type of argument that typically accompanies such documentation to establish that such reductions are unacceptable, a claim this chapter will label the "Unacceptability Thesis".

S. Koller (✉)
Faculty of Technology, Policy and Management, Section Philosophy/3TU.
Centre for Ethics and Technology, Delft University of Technology,
Delft, The Netherlands
e-mail: s.koller@tudelft.nl

C. Basta and S. Moroni (eds.), *Ethics, Design and Planning of the Built Environment*, Urban and Landscape Perspectives 12, DOI 10.1007/978-94-007-5246-7_4,

The remainder of the chapter addresses whether or not the reduction is indeed unacceptable from a philosophical point of view, in the sense of being theoretically false.

To do this, the chapter will first, in Sect. 4.3, try to support the Unacceptability Thesis by reference to Kant's writings on ethics and aesthetics. Outlining Kant's core ideas on how ethical and aesthetical value judgments are rationally supported will help us see that reducing one to the other is, indeed, philosophically infeasible. This concludes the third section.

In Sect. 4.4, the chapter tries to critically engage the Kantian argument just given and thereby cast doubt on the Unacceptability Thesis. Whether or not the Unacceptability Thesis is ultimately true, its credentials based on Kantian considerations are shown to be contestable.

Section 4.5 then attempts to interpret the original argument for the Unacceptability Thesis in a different vein, by not appealing to Kant at all and instead focusing on a naturalist reduction of ethical values to aesthetic properties, individuated naturalistically. This is also the first point that the term "reduction" receives some concentrated effort at disambiguation. For, it is argued, assessing whether or not ethics reduces to aesthetics requires preliminary clarification on the reduction relation at stake.

Section 4.6 concludes the chapter by casting doubts on the naturalist, property-oriented approach just outlined. The chapter does not, however, conclude with a firm verdict on whether architectural discourse is entitled to reduce matters of ethics to aesthetics or not. Rather, its aim will have been to single out the many theoretic issues that need to be addressed more fully for the basic issue of reduction to be resolved satisfactorily. By example, rather than exhortation, the chapter will have shown that certain areas of architectural discourse stands in need of philosophical regimentation to have any hope of progressing beyond the vagaries they incline towards, vagaries simply not up to the task for theoretic resolution.

Before we proceed, a preliminary observation. It may seem that the two concerns outlined in the opening of the chapter – to inquire *why* the field of architecture ethics is so underdeveloped and to attempt to *remedy* this lack of development – do not have immediate repercussions on one another. In that vein, inquiring why architectural theorists (outside philosophy) have attempted to reduce (their) ethics to aesthetics may help us little in the project to develop a philosophically feasible "ethics of architecture". However, Fisher (2000, p. 138) has suggested that a major reason why "architecture ethics" is so underdeveloped is precisely that architecture "has long been viewed as a branch of aesthetics rather than ethics. If anything, ethics has been thought of as applying to architects and not to architecture, to the actions of professionals, not the traits of buildings." It is not obvious (how) one can consistently maintain that, in architecture, ethics *reduces to* aesthetics and at the same times does not *apply* to architecture at all. But showing whether, and how exactly, the former is true or false could go some way towards removing that obstacle (the alleged inapplicability of ethics to architecture) on the road to developing an ethics of architecture.[1] And that is exactly the aim of this chapter.

4.2 Reduction of Values in Architecture: Modernism and Beyond[2]

Any recent history on the alleged reduction of ethics to aesthetics in architecture will sooner or later stumble across modernism, a movement that began to dominate architecture in the 1920s owing largely to its key figures Le Corbusier, Mies van der Rohe, and Gropius.[3] Modernist architects were not shy to issue grand claims and proclamations which intimately linked the ethical to the aesthetical dimension of architecture. Le Corbusier (2008, p. 254) for instance speaks of the mass production house (which modernism championed) as "healthy (morally, too) and beautiful from the aesthetic of the work tools that accompany our existence".

However, it took another three decades before these proclamations, and their impact on planning and architecture, had gained enough critical mass to attract countervailing voices. Manfredo Tafuri (1975, p. 178) remarked disparagingly about the "pathetic 'ethical' relaunchings of modern architecture", a verdict that was more fully borne out in Watkin's more systematic study *Morality and Architecture* (Watkin 1977). The stakes in voicing such countervailing concerns appeared high at the time, given how entrenched modernism was in both practice and theory, leading a contemporary reviewer of Watkin to comment: "In some countries Doctor Watkin would be well advised to hire a bodyguard or even to slip across the frontier! He deserves to be congratulated for standing up for intellectual freedom".[4]

However, with the demise of modernism in later decades, criticisms of its "ethics" have become comparatively commonplace, and "modernism" has come to serve a similar function to that of "Cartesian dualism" in philosophical circles, as a target which nearly everyone can agree on as hopelessly flawed and outdated. Here is such a contemporary summary, from Ockman (2009, p. 45):

> Among the main issues at stake [is] that of "good form" and the way modern architecture in particular proposed to see itself as a metaphor for, if not an actual instrument of, good society. Since the rise of modernism, the conflation of aesthetic values with ethical ones has been a reflexive habit in architectural thinking [...]. Invoking ideas like "structural honesty", "truth to materials", "good design" and "form follows function", architects from the nineteenth century on borrowed the mantle of morality to cloak or buttress value judgments based on aesthetic criteria or taste.

Studies like Ockman (2009), Leach (2005), or Till (2009) cement such summaries with ample reference to modernist writings, like the quotation from Le Corbusier submitted above. However, these studies go beyond making a historical case for the sins of a past movement and additionally attempt to show how the alleged "conflation of aesthetical values with ethical ones" remains a "reflexive habit in architectural thinking" to this day.

A much favoured example to press home this charge is the 2000 Venice Architecture Biennale, which its curator Massimiliano Fuksas launched under the telling motto "Less Aesthetics, More Ethics". As Till (2009, p. 175) observes, Fuksas' call to "address issues beyond the aesthetic" was "well-meant" if "fatally flawed because those four words still wedded aesthetics to ethics; they just asked for a rebalancing of the priorities". To show what is at stake here, Till walks us through the displays at that Biennale, such as Britain's (represented by David Chipperfield), and concludes (Till 2009, p. 175):

> The common message arising out of these voices [such as Chipperfield's] is simple: that ethics and aesthetics are mutually dependent; good aesthetics, in the form of beauty, leads directly to a good life, in the form of an ethical society, and equally that ethical society is the necessary context for the context of good aesthetics. This closed loop is very consoling for architects, because it places them—as arbiters of aesthetics—as central figures in the ethical process. The iteration of this loop was precisely the response of most architects to Fuksas's provocat[ing motto] at the Venice Biennale; not less aesthetics but actually more, on the understanding that as long as aesthetics can be equated with ethics, more aesthetics results in more ethics.

It will not be the purpose of this chapter to assess the factual accuracy of general diagnoses like these about architecture past or present. Rather, the focus will be on claims which characteristically accompany such diagnoses, claims to the effect that the alleged reduction of ethics to aesthetics in (and by) architecture is unacceptable. As stated, I will call this the Unacceptability Thesis. It finds its articulation to different degrees of conceptual sophistication, and for matters of convenience, we can focus on that offered in Till (2009, pp. 173–176). I regard terminological variants on "reduction" in Till's text, such as "equation with" in the quotation above, as inconsequential for these purposes.

On my analysis, Till wants to secure the Unacceptability Thesis by a *reductio* argument. A *reductio* is an argument that posits the negation of the statement (position) one wants to actually defend. By showing that negation to be untenable, one has indirectly vindicated the statement one wanted to actually defend. Here is the *reductio* I take Till to roughly press for[5]:

> *P1 Reified Ethics*: Architects engage moral values not (as these arise) in the realm of people but only in the realm of buildings (and construction elements) – more generally, in the realm of entities primarily individuated in terms of their aesthetic properties and primarily lacking in a social dimension.
>
> *P2 Phony Ethics*: Ethics has to be situated in the world of social dynamics. Once ethics is removed from that world, it becomes a "phony ethics".
>
> *C Conclusion*: Architects engage in phony ethics.

Leach (2005) levels a closely analogous argument at a more specific target, the writings of Karsten Harries. The "ethical function" of architecture *selon* Harries (1998) is to provide an *ethos*, a shared spirit that helps a human community find its place or identity in the world. That *ethos* is then primarily construed as a vehicle for characteristically idyllic modes of dwelling like rural farmsteads. It is not hard to see how this lands Harries with an "aestheticization" of ethics (to use a neologism of Leach's) and has his ethics of architecture fail "to engage substantively with social, economic, or political questions" that typically plague urban contexts (Leach 2005, p. 136). Whether or not this is a fair portrayal of Harries' position, the outline just given renders Leach's argument sufficiently parallel to Till's argument P1-P2-C so that whatever we may (and will) say of the merits and demerits of Till's argument can readily be applied to Leach's.

What, then, can we say about argument P1-P2-C trying to support the Unacceptability Thesis on which a reduction, in architecture, of ethical to aesthetic values is unacceptable?

To begin with, it appears that premise P2 is doing significant work. To see what is intended, consider Scruton's (1979, p. 320) claim that a well-made brick wall exhibits "honesty" and propriety which "embodies" a moral stance. This, according to Till (2009, p. 177), falls foul of "phony ethics", since "a brick has no morals". It is initially unclear whether this polemical response does justice to Scruton's claims. For one, Till has here exchanged Scruton's example of a "brick wall" with that of a single brick, when it is actually more intuitive to see the moral significance of the former, say, as a prison wall, than of the latter.[6]

However, this is where premise P1 becomes all important, for it urges us to individuate architectural entities without reference to their social dimension. A brick wall considered purely with reference to its aesthetic properties excludes reference to any social function it serves (a claim fully borne out in Kant's aesthetics, as we will see in Sect. 4.3). This highlights the fact of P2 doing "its work" to drive home conclusion C only in conjunction with premise P1.

What our analysis brings out so far is that Till's reasoning is logically valid in that his premises suffice to secure the intended conclusion C. Precisely because the *reductio* P1-P2-C *succeeds* and the conclusion C is deemed untenable ("an aesthetically underwritten ethics is no ethics at all"), Till has generated intellectual pressure to abandon at least one of the *reductio*'s premises (P1 or P2). If the premises entail an untenable conclusion, minimally one of the premises must go.

This explains why Till avails himself of a particular ontological understanding of architecture in P1 only to reject it all the more roundly later to present a position he *actually* favours. According to Till (2009, pp. 178ff.), a *proper* ontology of architecture will individuate architectural entities in a way that accommodates social relations and thus enables the possibility of "spatially empowering" architecture – for only *then* can architecture itself avoid committing itself to a "phony ethics". In other words, Till upholds P2 which expresses a constraint on what an "ethics" minimally has to be, but rejects P1. This is again consistent because premise P2 only secured an untenable conclusion *once conjoined to* P1. P2 by itself is a harmless premise, and not (deemed to be) inherently untenable like C.

Our analysis so far has carefully analysed the *logical structure* of Till's argument but has not treated the individual truth values of Till's premises P1 and P2. However, only if the premises of the argument *are* true is the argument sound too. We want to know which ontological understanding of architecture is the correct one – the one Till submits in P1 or the negation he champions later? We shall return to this towards the end of the chapter. For now, let us outline the consequences of Till's argument for the dialectical situation overall.[7]

It seems that whether or not architectural ethics reduces to aesthetics depends not even on how we individuate these two domains (architectural ethics, architectural aesthetics), but primarily (perhaps solely?) on how we individuate architectural entities. If we take these entities to be individuated, from the start, in inter alia social parameters, it seems then much harder to insulate architecture (thus construed) from such parameters and the ethical issues that such parameters characteristically engender. However, it is equally coherent to individuate architecture (i.e. architectural entities) insulated from social parameters and then to argue, consistently, that there is nothing inherent in architecture that merits ethical attention. The resulting position, discussed in the twentieth-century architectural theory under the label of "autonomism", admittedly attracted wide criticism. But even autonomism's fiercest critics would not allege the position to be incoherent or otherwise self-refuting.

Till's argument, then, does not succeed to substantiate the Unacceptability Thesis; it only highlights some consequences of the thesis which the subscribers of autonomism may or may not find unpalatable on pragmatic or moral grounds. However, there is something self-defeating in appealing to an amoralist's moral inclinations. Indeed, quite a few ethicists prefer telling moral sceptics to "get lost" to engaging them in substantive argument.

The remainder of my work will therefore tackle the Unacceptability Thesis on different grounds and abandon any further inquiry into Till's attempts to substantiate it. If we want to defend the thesis, we must approach it in a different vein.

In the next two sections, I will address the Unacceptability Thesis by recourse to epistemology. Based on Kant's writings, we can show how ethical and aesthetical value judgments differ too radically from each other to permit reduction of one to the other. Note that this is premised on a particular gloss on the Unacceptability Thesis which places it squarely within value theory. On this "gloss", reducing ethical *values* to aesthetic values is philosophically "unacceptable" because theoretically it is false. As we will see in Sect. 4.5, this is not the only way to construe the Unacceptability Thesis. It is, however, certainly a legitimate contender and captures the thesis as intended in the quotation from Ockman (2009, p. 45) we encountered in Sect. 4.2, on which "the conflation of aesthetic values with ethical ones has been a reflexive habit in architectural thinking".

4.3 Kant on Aesthetics and Ethics

There is a hope in contemporarily theorizing about values that if we pay particular attention to the epistemological issues involved in value ascription, we can start to see similarities in distinct types of values. For instance, Sturgeon writes:

> When philosophers have investigated the epistemology of morals – how, if at all, moral judgments can be justified, and how, if at all, there can be moral knowledge – they have usually thought that moral judgments, though distinctive, have a lot in common with evaluative and normative judgments in other areas, such as aesthetics. (Sturgeon 2006a, p. 242)

If Sturgeon were right, and we could (in addition) marginalize the "distinctness" of moral evaluative judgments compared to aesthetic ones, the Unacceptability Thesis would be threatened. My aim in this section is to show how this threat can be averted, if we accept Kant's accounts of the respective epistemologies of ethical and aesthetic value judgments.

To begin with, I follow Kant in simplifying discussion of aesthetical values to ascriptions of beauty. The role of "is beautiful" here is to serve either as an instance of aesthetic value or as a dummy predicate to express, at the most general level, a commendation on aesthetic grounds. (Contrast the predicate "is good", which signals commendation more generally.)

Further, my account assumes that Kant offers a normative, not a descriptive, account of beauty ascriptions. That is, Kant enumerates features he deems individually necessary and jointly sufficient to arrive at a *rationally defensible* judgment of something's being beautiful. If people on occasion (even frequently) arrive at aesthetic evaluations coming from a different route, that is both of no concern to him, nor does it (ipso facto) pose a counterexample to the account he gives. For the purposes of this work, I will delimit discussion of the aforementioned "features" of beauty judgments to two only, the focus on representations rather than objects and the notion of "disinterest" to spell out shortly.

To understand the first of these "features", consider the opening lines of Kant's major work on aesthetics, his late *Critique of the Power of Judgment*[8]:

> The judgment of taste is aesthetic: In order to decide whether or not something is beautiful, we do not relate the representation by means of understanding to the object for cognition, but rather relate it by means of the imagination (perhaps combined with the understanding) to the subject and its feeling of pleasure or displeasure. The judgment of taste is therefore not a cognitive judgment. (Kant, AA V:203, tr. Guyer 2000)

The first "feature" to single out in this account is that, to judge an object beautiful, we attend to the quality of pleasure or displeasure as it arises from the *representation* of the object. Kant explains the prerequisite notion of "representation" in his *Metaphysics of Morals*, when he writes:

> The capacity for taking pleasure or displeasure in a representation is called feeling because both of these involve what is merely subjective in the relation of our representation and contain no relation at all to an object for possible knowledge of it (or even knowledge of our own condition). (Kant, AA VI:211–212, tr. McGregor 1996b)

Once conjoined with the previous passage, this immediately rules out that judgments of beauty can ever attain the status of objective knowledge. This generates a problem for the "value reductivist", who holds a reduction of ethical to aesthetical values feasible, as follows.

As Kant points out in his *Groundwork to the Metaphysics of Morals*, moral value judgments have to carry sufficient deontic weight to command their ascent by any rational, human subject. This, however, has the immediate consequence that such value judgments cannot be underwritten by (and hence, depend for their validity on) the specific psychological "make up" of particular human subjects, as that "make up" need not (and cannot be rationally commanded to be) shared by all rational agents. As he remarks,

> moral precepts should not be valid only under the contingent conditions of humanity [, ... for] how should laws of determination of *our* will be taken as laws of the determination of the will of rational beings as such, and for ours only as rational beings, if they were merely empirical [?]. (Kant, AA IV:408, tr. McGregor 1996a)

This principle leads Kant to exclude moral values and precepts to be contingent on specific human emotions, desires, and so on. Which is, as we just saw, he premised his account of aesthetic value judgments on. And that is the first problem for the value reductivist.

To be sure, Kant recognizes intersections between ethics and aesthetics (see his *Critique of the Power of Judgment*, §60), but at no point is he inclined to permit a reduction of one to the other. And his primary reason for that, as we just saw, is that he thought he could ground ethics on rational "necessities", whereas aesthetics was inextricably bound up in the contingencies of the human-sensitive apparatus. I mention this because Julian Roberts (2005, p. 153) has recently argued that a Kantian aesthetics is unsuitable for architecture since that aesthetics is overly "cognitive, generalizing, and 'scientific'", whereas "[o]f all the arts, architecture resists cognitive analysis most strongly" ("buildings do not open themselves to strenuous contemplation" but rather "work at an intuitive, sensory level"). While Roberts raises a legitimate worry – the potential unsuitability of a Kantian aesthetics as regards architecture – it should be clear that attention to Kant's text does not bear out its characterization in Roberts. (Nor is it clear Roberts does justice to the aesthetic appraisal of architecture, but I leave that for others to decide.)

The second "feature" of judgments of beauty *selon* Kant salient to assessing the value reductivist's claim is what Kant calls the "disinterest" of aesthetic value judgments:

> The satisfaction that determines the judgment of taste is without any interest. (Kant AA V:204, tr. Guyer 2000)

This is to be explained as follows. In attending to an object, that is, its representation, aesthetically, we "bracket out" all aspects that are irrelevant for appreciating it aesthetically. Here is an instructive example:

> Suppose we have an image of a palace before us. To deny that the palace is beautiful because it took an excessive amount of money and human labor to build it, would confuse the cause of the building with how the building looks. (Wicks 2007, pp. 21–22)

Imagine you see a postcard of the pyramids in Giza (or, even luckier, you have the privilege of standing right in front of them), and deem them beautiful. (In this use of the example I diverge from Kant who, in § 26 of his *Critique of the Power of Judgment*, relates the pyramids to the sublime rather than the beautiful.) You are not supposed to withhold your initial aesthetic appraisal once you are told how many slaves probably had to die to build these monuments. And the reason for that relates to the first "feature" we enumerated in Kant's account, according to which one exclusively attends to a thing's surface appearance to you, its sensual "representation". You attend to how it *looks* and are supposed to ignore all the background story and background knowledge you may have about the thing when you assess it aesthetically.

Pretty much *any* argument to the effect that it is philosophically viable to reduce ethical to aesthetical value judgments got to be over at this point. To say that we can basically forego ethics because aesthetics will take care of ethics lands us with praising the pyramids and giving thumbs up to slave labour.

However, this potentially overestimates the impact of Kant's remarks a propos "disinterest". For, it could be said that any moral indignation about slave labour attaches at best, not to the object whose aesthetic properties are at stake (the pyramids), but to the historical process which gave rise to it and, so, is causally distinct from the pyramids. It is not as if the original value judgments had claimed that the causal process was beautiful, only that what it gave rise to was.

Whatever we ultimately make of Kant's notion of "disinterest", then, it should be clear by now how, if Kant is correct, the way to arrive at ethical and aesthetical value judgments in a rationally defensible way radically diverges. Observe how, to drive this conclusion home, we did not have to rely on many specifics of his accounts. For instance, his claim to ground morally normative judgments in so-called categorical imperatives (formulated in his *Groundwork*, AA IV:421) did not even need to be mentioned. I regard that a strength of the position in that it requires fewer premises than expected.

That said, the whole counterargument to the value reductivist's claim, and (thereby) indirect support for the Unacceptability Thesis, on Kantian grounds most definitely relies on the philosophical correctness of these "grounds", however cursorily specified.

Several objections come to mind. As regards Kant's ethics, and the demand for it to ignore the contingencies of the human "psychological makeup", Irwin (2009) has recently observed how Kant, writing before Kripke's (1980) discovery of the *a posteriori necessary*, fallaciously conflates "the empirical" (to which human nature and psychology belongs) with "the contingent". If Kripke is right, empirical reality is not devoid of necessities. Kant's primary criterion to set ethics apart from aesthetics, thereby, is under threat.

Further, the correctness of Kant's aesthetics assumes, rather than argues for, the incorrectness of a realism about aesthetic experience, a position we can define as follows:

> a[n aesthetic] realist might say that aesthetic experience is experience that is endowed with aesthetic representational content. This means that our aesthetic experience represents aesthetic states of affairs, situations, or facts. This, in turn, means that in aesthetic experience the world is represented as possessing genuine aesthetic properties. Such experiences ground or rationally cause our aesthetic judgments, which also have such realistic representational content. (Zangwill 2003, p. 64)

A defence of the Unacceptability Thesis on Kantian grounds, therefore, would need to be carefully weighed against these objections. However, the truth of aesthetic realism (if that truth could be established) would potentially not just undermine the strategy pursued in this section, oriented on Kant's work, but to any attempt to defend the Unacceptability Thesis on epistemological grounds. The next section explains why.

4.4 A Problem for Kant: The "Fregean Rejoinder"

The Kantian strategy just pursued can be stated more generally, without *any* (not even "cursory") reference to the specifics of Kant's value epistemologies in ethics and aesthetics. Assume two distinct predicates "F" and "G", such as (but not exhaustively) "beautiful" and "morally good". Then the generalized Kantian strategy is to argue that "F" and "G" pick out distinct (mutually irreducible) properties because how we rationally ought to arrive at judgments of "x is F" differs from how we rationally ought to arrive at judgments of "x is G". What "generalizes" the Kantian strategy here is that "how one rationally ought to arrive at" judgments of a particular variety is not spelled out. It is not spelled out because, regardless of how it is specified, the following objection still holds. I call it the "Fregean Rejoinder", in honour of Gottlob Frege's 1892 paper on identity (Frege 1994), which focuses on the (self-) identity of the planet Venus picked out by the descriptions "the evening star" and "the morning star":

> *Fregean Rejoinder:* epistemic distinctness does not entail ontological distinctness. The first engages the realm of "sense", the latter the realm of "reference". We might *think* of the evening and the morning star as distinct existences (due to the different representations, or "senses", we have of them), but this does not mean that they are distinct existences. To the contrary, they are one and the same planet – Venus.

Similarly, just because our "epistemic routes" to aesthetic and ethical values are distinct does not mean that the values themselves are distinct. You may see, touch, and hear a certain physical subject, say, a person; you hear them talk, you can shake their hand, and you can see them. It would be mistake to infer from the disagreeing deliverances of these sensory modalities that you are experientially related to three persons, not one.[9]

However, the force of the Fregean Rejoinder is less clear than it might appear at first sight. To be sure, as just stated ("three persons, not one"), it sounds compelling. But that is on the assumption that what sensory modalities pick out are persons and not their sensory properties. Certainly we are talking of the same person, but are we not also talking about distinct properties – the person's look, the person's smell? Surely *those* are distinct properties. Similarly, it is not as if our value judgments could in any comprehensible sense point us to the same *referent* – rather, it seems as if they pick out distinct *properties*. And that is what we are after – not whether one and the same thing can *have* both ethical and aesthetic properties, but *whether the properties thus had* are one and the same in given instances.

Even this objection to the Fregean Rejoinder loses force once scrutinized, for there are properties which are accessible to several sensory modalities. For instance, I can both see and touch the spatial properties of that table in front of me. I would be mistaken to conclude that the table possesses two distinct sets of spatial properties, even though my senses arguably disclose two ways of experiencing them. And that is the Fregean Rejoinder all over: epistemic distinctness (here, distinct experiences of the same properties) does not entail distinctness of properties. By the same token we may observe how, just because one experiences two properties as being the same does not mean that they thereby *are* one and the same property. Which is to say, the phenomenology of our moral and aesthetic emotions could be severely misleading. It could represent two properties as being one, and one as being two.[10]

To summarize the current section, what the "Fregean Rejoinder" highlights is that one cannot defend the Unacceptability Thesis on purely epistemological grounds, regardless of which particular "value epistemology" one adopts. The force of this objection can only be averted at great cost. For instance, one could claim (as Kant, incidentally did) that metaphysics and epistemology do not "come apart", are not ultimately distinct, and, hence, any objection that rests on their distinctness, as the Fregean Rejoinder does, is mistaken. But to argue for this "non-distinctness" is a heavy burden and an assessment of its viability well beyond the scope of the current work. Suffice it to say that moving back to a position which integrates epistemology closely with metaphysics hearkens back to the days of Kantian and post-Kantian idealism, a position that mainstream philosophy has moved beyond a good hundred years ago.[11] To resuscitate it in the name of arguing for the distinctness of ethics from aesthetics seems disproportionate. Luckily, the Unacceptability Thesis does not have to put all its eggs in that one (Kantian) basket. As the next section shows, a more promising defence is available, one associated with considerably lesser philosophical "costs".

4.5 An Alternative to Kant: Naturalism and Property Reductionism

Recall the overarching agenda of this chapter: to defend the Unacceptability Thesis on which a reduction of ethics to aesthetics is theoretically untenable, whether in architecture or elsewhere. I glossed this claim in terms congenial to value theory, as the unacceptability of reducing ethical values to aesthetical ones. I then looked at Kant's value epistemologies to conclude that to sustain the Unacceptability Thesis on such grounds is less clear cut than it initially seemed. What we found out in the last section, in particular, was that no matter how much we cut down on specific premises and commitments to spell out these "grounds", insurmountable difficulties loom. This may cause us to retread our steps to the opening move and query whether glossing the Unacceptability Thesis in value theoretic terms was the most promising avenue.

Once we take this "step back" from our original formulation, we can actually see how Till's own words and even the regimented argument I presented on his behalf considerably underdetermine whether we ought to cast the Unacceptability Thesis in terms of aesthetic *values* at all. For, and this is the key point, Till may (on behalf of the architects he castigates as inclining to a "phony ethics") have in mind the reduction of ethics, even ethical values, to aesthetic *properties*, nonevaluative ones. Instead of having in mind the type of aesthetic values, or evaluative properties, that, in Sect. 4.3, we gathered under the dummy predicate "is beautiful", Till may have in mind a reduction to properties relevant to aesthetic appraisal, aesthetic properties in the sense of a thing's shape, colour, or configuration. (The very type of non-evaluative properties Kant's account of aesthetic appraisal worked on.)

This single move, from values to nonevaluative properties, considerably alters our theoretic options. Also, we may now observe how the very reduction relation itself, of one type of properties to another (whether evaluative or not), is itself in need of greater clarification.[12] Metaphysical work in the philosophy of mind has shown that we can choose from a large number of theoretic accounts to construe property correlations, only some of which would ultimately be deemed "reductive" in a strong sense (a sense to merit, on pragmatic grounds, its "unacceptability"). Instead of enumerating the (vast) possibilities at our disposal here, I will focus on one particular instance which has recently been proposed by Sydney Shoemaker, with an eye on (inter alia) architectural properties. Shoemaker's starting point is "physicalism", a position on which "all states and properties of things, of whatever kind, are physical or physically realized" (Shoemaker 2007, p. 1). He states:

> I think that mental properties are physical in the same sense that automotive properties, *architectural properties*, computer properties, and botanical properties are physical. I would express this sense by saying that instances of all such properties are physically realized—they are [...] property-realized in accordance with my subset account. (Shoemaker 2010, p. 125, emphasis added)

What Shoemaker refers to (here) as his "subset account" has two prerequisites, the first of which, following Armstrong (1997, p. 41), we may call the "Eleatic Principle":

> *Eleatic Principle*: Only those things which are "causally active" exist.

This principle exerts a causal constraint on which properties *exist*, in that causally idle properties are deemed to be non-existent. Shoemaker goes a step further and coins a principle which exerts a causal constraint on the *individuation* of properties:

> *Causal Individuation of Properties*: "what makes a property the property it is, what determines its identity, is its potential for contributing to the causal power of the things that have it". (Shoemaker 1984, p. 212)

This principle allows Shoemaker (2007) to formulate the claim of one (type of) property being "realized" by another (type of) property in terms of what he calls "my subset account" in the quotation provided above.

> *Subset Account*: Property F realizes property G if and only if the causal powers of G are a proper subset of the causal powers of F.[13]

Shoemaker's claim as regards architectural properties (referenced in my quotation from his 2010) can now be explained as follows. Architectural properties like shape and configuration (e.g. the arrangement of one brick on top of another) are "realized" by physical properties, in that the causal features making up the identity of architectural properties form subsets of the causal features of physical properties which realize them.

The upshot of this claim for our debate is as follows. If ethical values "reduce to" architectural properties such as shape and configuration, in a sense of "reduce to" to be yet made precise, and these properties are *realized by* physical properties in the sense specified, then we can assess the merits of reducing ethical values to non-evaluative, aesthetical properties (such as "architectural properties") without paying any attention to the "middle man" in that argument. We can, basically, leave out all reference to architectural properties and simply evaluate the cogency of reducing ethical values to physical or (more generally) naturalized properties. And this significantly improves our prospects to philosophically evaluate the cogency of the reduction of ethics to "architectural properties", for this reformulation casts the reduction claim as a variety of meta-ethical naturalism, a position whose cogency is comparatively well researched.

This is the first theoretic advantage over the original (value-theoretic) formulation of the claim rejected by the Unacceptability Thesis. The second advantage is that the literature on "realization" shows that the subset account, in particular, affords us property correlations contingent on the laws of nature that hold in the actual world. This is because the very features which make up the "identity" of the properties thus correlated are contingent on the laws of nature in the worlds in which the properties are instantiated. But this means that the "nomic" force of realization claims only covers possible worlds which share their laws of nature with the actual world and does not "fan out" across all possible worlds, including those not sharing our laws of nature. Since, however, only property correlation claims which are strictly necessary, and "fan out" across all possible worlds, are deemed "strongly reductionist", it follows that realization claims are not reductionist in any particularly worrying or "unacceptable" manner.[14] And this, finally, has the consequence that *if* architects would gloss their claims of reducing ethical values to aesthetic ("architectural") properties in Shoemaker's terms, *then* they would underpin their claim by a theory that looks philosophically robust and averts the charges of pragmatic "unacceptability" that underpins strongly reductionist varieties of property correlation.

4.6 Problems for the Realization Strategy

As with the Kantian strategy pursued in Sects. 4.3 and 4.4, it behoves us to take stock not only of the current strategy's theoretic benefits but also of its drawbacks, for, while comparatively robust, the strategy makes a couple of theoretic moves that are certainly open to philosophical objection. Let us go through these one by one.

To begin with, the very first step, the "Eleatic Principle" has come under fire for eliminating a high variety of so-called abstract entities from our ontology (from the set of things, we are committed to believe there are). It is unclear whether numbers can meaningfully be said to be causally efficacious, for instance, and it is unclear whether the "Eleatic Principle" can respect that fact without compromising huge swaths of the natural sciences. Field (1980), for instance, himself committed to the Eleatic Principle, attempted to recast mathematics as "a science without numbers". It remains hotly debated whether this attempt is ultimately tenable.

More particularly, the question arises whether the very properties we want to reduce, ethical ones, can be meaningfully said to be causally efficacious in the sense demanded by the Eleatic Principle and the Causal Individuation of Properties. This, too, is a topic philosophers are divided about. Sturgeon (2006b, p. 100) certainly believes

> we can appeal to the apparent causal role of ethical properties in the natural order. Common sense agrees with a long tradition of philosophical thought in assigning ethical properties such a role. Most of us can identify occasions on which we think we have benefited from someone else's goodness or been harmed by their moral faults.

However, as Dancy (2006, p. 127) shows, this observation can be easily challenged. Instead of arguing for the causal efficacy of ethical properties (naturalized or otherwise), the observational data are consistent with the causal efficacy being exercised by the beliefs, and consequently actions, of human agents thinking about (states of affairs involving) these properties, without these properties entering a causal chain of any sorts. Dancy (ibid.) continues:

> As for the example of benefiting from the goodness of others, the response might be that what we benefit from are those features of others that make them good, for instance their concern for their fellows, or their willingness to put themselves second. It is the good-making features that are affecting the causal order, not the goodness that they make.

However, Dancy's counterargument against applying the Eleatic Principle and the Causal Individuation of Properties to ethical properties at this point assumes, rather than argues for, the incorrectness of applying the "subset account" to explain the relation between ethical properties and what Dancy here calls their "good-making features". For, on the subset account, the respective causal features of these two do not stand in competition, are not the causal features of *one rather than the other*, since it is precisely (an overlap of) the very same features enjoyed by both (see Shoemaker 2001, on how the causal efficacy of the realizer does not "pre-empt" that of what it realizes).

This means that assessment of the "realization" strategy to undermine the Unacceptability Thesis squarely rests on the cogency of the claim that we can relate ethical properties to nonevaluative ones by way of the subset account. Sturgeon suggests how this claim can be sustained:

> If a naturalist really wants to use the identification of heat with molecular motion (for example) as a model for a reductive account of some ethical property, then it is worth noting that the grounds for the scientific identification lie largely in a matching of causal roles. (Sturgeon 2006b, p.100)

The problem with the position suggested here, we can now see, is that it involves an appeal to meta-ethical naturalism, a view on which ethical properties "reduce to" natural ones. And this is commonly deemed to be deeply problematic, to rest on an "is/ought confusion", and to be untenable for reasons Moore (1993) showed a century ago. However, as Sinnott-Armstrong (2000) has recently argued, Moore's arguments ought to be re-evaluated. For instance, it is unclear whether no "is" could ever entail an "ought", in that (for instance) "Bertie and Madeleine are dead" might entail "It is not the case that Bertie ought to marry Madeleine".

This is not the place to evaluate the cogency of the case for and against meta-ethical naturalism. However, as long as the case against is not conclusively settled, Sturgeon is entitled to say, as he does (2006b, p. 92), that "such ethical properties as the goodness of persons […] and such as the rightness or wrongness of actions, are natural properties of the same general sort investigated by the sciences". And nothing more is required for the subset account to go through against the objections levelled against it.

However, there is one final worry that is significantly harder to dispel. This whole section has been squarely oriented on matters of *ontological* reduction (if reduction in a "soft", philosophically "acceptable" sense, that of "realization"). Even if all the ontological worries against that strategy could be shelved, we would be none the wiser about the suggested reduction, here, of ethical values to architectural properties, being acceptable at the *conceptual* level or not. The following observation applies to any (however weakly reductionist) variety of property correlation between two property types F and G, no matter how we terminologically label it and theoretically spell it out ("the Fs are realized by the Gs", "the Fs are grounded in the Gs", "the Fs supervene on the Gs"):

> However long a list we give of the items to which a supervening term [F] applies, described in terms of the level supervened upon [G], there may be no way, expressible at the level supervened upon [G], of grouping just such items together. [...] Understanding why just those terms belong together may essentially require understanding the supervening term. (McDowell 1998, p. 202)

The worry issued here is that no matter how refined and ontologically acceptable a reduction of one set of properties to another may be, we may still wish people to engage in the original type of properties (F), not (only) the reduction base. In the case at hand, we may still wish architects to engage matters of human ethics "hands on", even if (which is not clear) such matters ontologically reduce to areas architects are more comfortable to handle and better trained at handling (the aesthetic properties of construction materials and their like).

The moral, we may say, here is the mirror image of the moral held in Sect. 4.4. Whereas the "Fregean Rejoinder" showed us that epistemology is no substitute for ontological analysis, the final worry of this section highlights that ontological analysis may not be suited to dispense all our epistemological worries about reduction, including worries about understanding the properties to be reduced.

It may also be said that, with all fairness, this final "worry" perhaps best expresses the concerns raised by Till and others by voicing the Unacceptability Thesis in the first place. What makes reducing ethics to something else so "unacceptable" is that it has people stop to *engage* with ethics, a fact that is deemed to have disastrous pragmatic *consequences* regardless of how theoretically innocuous the reducing may be in its own right.

However clear that may sound, it is, once more, an issue that loses its force upon closer scrutiny. As Crane (1999, p. 27) writes,

> It is sometimes said that a reductive identity theory [which reduces mental to naturalized phenomena] denies the existence of mind; but this is a simple mistake. To identify phenomena A and B is to deny neither the existence of A nor of B; on the contrary, the identification presupposes their existence. Despite this, many object that the kind of explanation given by reductionists must inevitably "leave something out". Expressed in this way, the objection is fatuous. The mechanical reductive explanation of thermodynamic properties is a genuine explanation: that is, it is an advance in our understanding of the phenomena explained. If there were a parallel explanation of mental phenomena in neuroscientific terms, then this too would be an advance in our knowledge. The mere fact that, if such an explanation were provided, it would be reductive, should not be a reason for rejecting it.

Crane here highlights a fact we may (re)state as follows. However brief the time required to state (and defend) the general case of reducing ethics to aesthetical properties may be, to sustain that case at an individual level, requires considerable work. It requires nothing less than the matching of causal features along a variety of scenarios, on a huge spectrum of ethical and architectural properties. If this "matching" could be achieved at the individual, concrete level, and not just generically defended on purely theoretic grounds, then our understanding of the ethics of architecture would actually be vastly enriched, not impoverished, by the reduction. (Here, as before, Till et al. may have tacitly construed the reduction relation as unduly eliminativist.) And, consequently, the "disastrous pragmatic consequences" Till and others may allege the reduction to have would not necessarily come to pass. Thus, of all the worries that beset a "realization strategy" to undermine the Unacceptability Thesis, worries of conceptual impoverishment are perhaps the least decisive.

The real worry, and with this I conclude, may well be that the ontology of buildings suggested here on behalf of the architect, that of (ultimately) physical properties, could be rejected by architects as inadequate and as not capturing how they engage with their work in the first place.[15] This objection highlights a key difficulty with evaluating the cogency of various "reductivist" strategies at the disposal of the architect: these strategies will sooner or later saddle him with a tacit ontology of architecture. Recall for instance how, in Sect. 4.2, Till availed himself of a particular ontology of architecture in his premise P1, only to reject it all the more roundly later. It seems, therefore, that an ethics of architecture requires as a preliminary a better theoretic grasp of its ontology. But like so many other issues touched upon in this chapter, this must await resolution on other occasions. If (methodologically speaking) this chapter is on the right track, such resolution will require a joint effort from architecture and philosophy.

Notes

1. "Some way" rather than "all the way", since the prospects of an ethics of architecture are beset by many additional "obstacles", some of which are outlined in Illies and Ray (2009, pp. 1127–1129).
2. For purposes of argument, this section follows Till's (2009) presentation of architectural modernism as a theoretically unified position. However, as Mallgrave (2005, pp. 261–271) has recently shown, a proper understanding of architectural modernism needs to take account of its theoretical diversity and historical breadth. Cautious readers may therefore wish to replace occurrences of 'architectural modernism' in the chapter with 'architectural modernism, in one of its guises' or 'architectural modernism, in some of its moments': the end result retains the chapter's ambition to afford a respectable inquiry into some of architectural modernism's theoretical underpinnings.
3. I do not here mean to deny any traces of such reduction in earlier architectural writers like Vitruvius (on who see Leach 2005, pp. 135–136n.3). However, bringing in such earlier writers into the purview of this chapter would needlessly complicate its agenda, since premodern authors, including architectural ones, were probably accustomed to a broader understanding of "the ethical" than modern ones (Williams 1986). I hope to rectify this limitation on future occasions.
4. This comment is reprinted on the blurb of the 2003 reprint of Watkin (1977).

5. I say "roughly" because Till himself does not present his argument in this regimented manner.
6. I owe this observation to Pieter Vermaas.
7. A full truth evaluation of Till's premises are beyond the scope of this chapter. Viewed against the larger literature, however, it is fairly easy to show that Till's premises are far from controversial. Premise P1 raises the proper ontological individuation of architecture. Kroes and Meijers (2006) have urged to adopt a so-called "dual nature" view of artefacts on which they are to be individuated both as physical objects with a function and as having a "social dimension". Whether or not it is correct to think of buildings as artefacts, if we can apply this "dual nature" individuation to architecture (as Primus and Kroes 2008, have claimed), this would immediately rule out premise P1. Going now to the other premise, Till's P2 highlights in what sense Till would reject the general term "ethics of the built environment" which has come to create its own subdiscipline in the past years (see, e.g. Fox 2000 and Fewings 2008). Till would not so much claim that the very term is a misnomer, since there only could be an ethics of people, not buildings (or landscapes). However, the moment such an "ethics of the built environment" stops to factor in, at square one, the very population of people in those environments, it has stopped to merit the term "ethics" in the first place. This generates an interesting dilemma for the very legitimacy of an "ethics of the built environment", if Fox (2009) is right to claim that its legitimacy depends on addressing concerns not already covered by extant ethics. According to Fox's argument, concerns relating to forms of professional conduct, physical, and psychological impact of architecture on building users can be absorbed into extant (meta) ethical frameworks dealing with human well-being, whereas issues concerning the "symbolic" and physical impact of buildings on the natural environment can be absorbed by extant ethics of the *natural* environment. The only remaining, and not thus "absorbable", area for an ethics of the built environment to address would be "a building's 'design fit', that is, the extent to which a building fits with its natural, social and built contexts when considered purely in terms of its design" (Fox 2009, p. 389). But then, the only sense for an ethics of architecture or the built environment to enjoy legitimacy *selon* Fox is for it to be a "phony" ethics in Till's sense (P2). And that is a straightforward contradiction. The desideratum to avoid the contradiction creates considerable pressure to either reject Fox's argument or Till's premise P2.
8. As is customary, references to Kant are to volumes of the *Akademie-Ausgabe* (AA) of Kant. Translations used are referenced at the end of this chapter.
9. One can express that Fregean Rejoinder at either the level of reference (as I just did) or even at the level of senses, provided one accepts that these senses are not "luminous" in Williamson's (2000, p. 95) sense. Basically, a condition C is luminous if, for any case, in that case a condition C obtains, one is in a position to know that it obtains. For instance, if one is in pain or feels cold, then one is in a position to know that one is in pain or feels cold. Feeling pain or feeling cold is "luminous" in that sense. If the copresence, sameness, and distinctness of aesthetical and ethical values, when "experienced", are not luminous, the Fregean Rejoinder can be restated at the level of sense.
10. I have here glossed over many issues, the most crucial one, the question on which properties' experience can be said to represent at all (cf. Siegel 2009). According to moral realism as stated earlier, aesthetic properties can be experienced. But this is an assumption that needs to be defended, not simply stated. However, if Price's (2006) 'sparse' view is correct, experience represents neither aesthetic nor ethical properties, and their parallel (employed in my discussion) to *spatial properties represented in experience* becomes rather loose. That said, we can (re)formulate the Fregean Rejoinder by recourse to *which properties our value judgments refer to* rather than *which properties experience represents*, and retain its full cogency as an objection to the Kantian strategy.
11. For a systematic assessment of the viability of that "integration", see Peacocke (1999).
12. The current point is how Sects. 4.2 and 4.3 have potentially misconstrued both the extension of the G's and the intention of "reduces to" in "the Fs reduce to the Gs" (where "the Fs" are ethical values pertinent to architecture).

13. The notion of "proper subset" is used here rather than that of "subset" *simpliciter* to rule out "x realizes y" being reflexive and symmetric.
14. I owe the argument on the nomic force of "x realizes y" to Kim (1998, pp. 22–23) and Kim (2004, p. 572). The observation on which degree of "nomic force" qualifies as "strongly reductionist" is owed to Blackburn (1993, pp. 61–62), who relates the debate to the nature of ethical properties.
15. Thanks to Andrej Radman for raising this objection.

Acknowledgments I thank the participants of the Delft Philosophy Colloquium for valuable feedback on an earlier version of this work and Andrej Radman and Sabine Roeser for written comments. Maarten Franssen, Bjørn Jespersen, and Peter Kroes have done much to improve (parts of) later drafts, particularly Sect. 4.2.

References

Armstrong DM (1997) A world of states of affairs. Cambridge University Press, Cambridge
Blackburn S (1993) Supervenience revisited. In: Blackburn S (ed) Essays in quasi-realism. Oxford University Press, Oxford, pp 130–148
Corbusier L (2008) Towards an architecture. Frances Lincoln, London
Crane T (1999) Review of Jaegwon Kim's mind in a physical world. Times literary supplement, August, (Issue 5029), p 27
Dancy J (2006) Nonnaturalism. In: Copp D (ed) Oxford handbook of ethical thought. Oxford University Press, Oxford, pp 122–145
Fewings P (2008) Ethics for the built environment. Routledge/Taylor and Francis, London
Field H (1980) Science without numbers. Blackwell, Oxford
Fisher TR (2000) In the scheme of things: alternative thinking on the practice of architecture. University of Minnesota Press, Minneapolis
Fox W (2000) Ethics and the built environment. Routledge/Taylor and Francis, London
Fox W (2009) Architecture ethics. In: Olsen JKB et al (eds) A companion to the philosophy of technology. Blackwell, Oxford, pp 387–391
Frege G (1994) Über Sinn und Bedeutung. In: Patzig G (ed) Gottlob Frege: Funktion, Begriff, Bedeutung – Fünf logische Studien. Verlag Vandenhoeck & Ruprecht, Göttingen, pp 40–65
Harries K (1998) The ethical function of architecture. MIT Press, Cambridge, MA
Illies C, Ray N (2009) Philosophy of architecture. In: Meijers A (ed) Handbook of the philosophy of science, vol 9, Philosophy of technology and engineering sciences. Elsevier, Oxford, pp 1199–1256
Irwin T (2009) The development of ethics: a historical study, vol 3: From Kant to Rawls. Oxford University Press, Oxford
Kant I (1996a) Groundwork of the metaphysics of morals. In: McGregor M (ed and trans) Kant. Practical philosophy. Cambridge University Press, Cambridge, pp 37–106
Kant I (1996b) Metaphysics of morals. In: McGregor M (ed and trans) Kant. Practical philosophy. Cambridge University Press, Cambridge, pp 353–600
Kant I (2000) Critique of the power of judgment. Guyer P (ed and trans) Cambridge University Press, Cambridge
Kim J (1998) Mind in a physical world. MIT Press, Cambridge, MA
Kim J (2004) Supervenience, emergence, realization, reduction. In: Zimmerman D (ed) The Oxford handbook of metaphysics. Oxford University Press, Oxford, pp 556–584
Kripke S (1980) Naming and necessity. Blackwell, Oxford
Kroes P, Meijers A (2006) The dual nature of technical artifacts. Stud Hist Philos Sci 37:1–4
Leach N (2005) Less aesthetics, more ethics. In: Ray N (ed) Architecture and its ethical dilemmas. Routledge/Taylor & Francis, London, pp 135–142

McDowell J (1998) Mind, value, and reality. Harvard University Press, Harvard
Mallgrave HF (2005) Modern architectural theory, a historical survey 1673–1968. Cambridge University Press, Cambridge
Moore GE (1993) Principia ethica. Cambridge University Press, Cambridge
Ockman J (2009) Ethics and aesthetics after modernism and postmodernism. In: Iliescu S (ed) The hand and the soul. University of Virginia Press, Charleslottville/London, pp 44–75
Peacocke C (1999) Being known. Oxford University Press, Oxford
Price R (2006) The way things look. Doctorate thesis, University of Oxford. Copy available for inspection from the Bodleian Library, University of Oxford
Priemus H, Kroes P (2008) Technical artifacts as physical and social constructions: the case of Citéde la Muette. Housing Stud 23:717–736
Roberts J (2005) Architecture, morality and taste. In: Ray N (ed) Architecture and its ethical dilemmas. Routledge/Taylor and Francis, London, pp 143–154
Scruton R (1979) The aesthetics of architecture. Methuen, London
Scruton R (1995) The classical vernacular: architectural principles in an age of nihilism. Macmillan, London
Shoemaker S (1984) Causality and properties. In: Shoemaker S (ed) Identity, cause, and mind. Cambridge University Press, Cambridge, pp 206–233
Shoemaker S (2001) Realization and mental causation. In: Gillet C, Loewer B (eds) Physicalism and its discontents. Cambridge University Press, Cambridge, pp 74–98
Shoemaker S (2007) Physical realization. Oxford University Press, Oxford
Shoemaker S (2010) Reply to my critics. Philos Stud 148:125–132
Siegel S (2009) Which properties are represented in perception? In: Szabó Gendler T, Hawthorne J (eds) Perceptual experience. Oxford University Press, Oxford, pp 481–503
Sinnott-Armstrong W (2000) From "is" to "ought" in moral epistemology. Argumentation 14:159–174
Sturgeon NL (2006a) Moral explanations defended. In: Dreier J (ed) Contemporary debates in moral theory. Wiley-Blackwell, Oxford, pp 241–262
Sturgeon NL (2006b) Ethical naturalism. In: Copp D (ed) Oxford handbook of ethical thought. Oxford University Press, Oxford
Tafuri M (1975) Architecture and utopia: design and capitalist development. MIT Press, Cambridge, MA
Till J (2009) Architecture depends. MIT Press, Cambridge, MA
Watkin D (1977) Morality, and architecture: the development of a theme in architectural history and theory from the gothic revival to the modern movement. University of Chicago Press, Chicago
Wicks R (2007) Kant on judgment. Routledge/Taylor and Francis, London
Williams B (1986) Ethics, and the limits of philosophy. Harvard University Press, Harvard
Williamson T (2000) Knowledge, and its limits. Oxford University Press, Oxford
Zangwill N (2003) Aesthetic realism 1. In: Levinson J (ed) The Oxford handbook to aesthetics. Oxford University Press, Oxford, pp 63–79

Part II
Evolving Paradigms

Chapter 5
Architecture as an Object of Research: Incorporating Ethical Questions in Design Thinking

Lara Schrijver

5.1 Introduction

This contribution presents architecture as a complex activity with no optimal design solution. Architecture and the built environment form a complex whole, consisting of many different elements serving various functions. This precludes the possibility of finding a single optimal solution. Thus, the problem of architectural design typically revolves around normative choices and a response to typically contradictory demands. By introducing descriptive ethics (what is appropriate to the built environment, which "moral beliefs" are embedded in it?) and normative ethics (how do we propose it is best to live or to address our environmental issues?), we can introduce a higher precision in design decisions.

When considered in this light, the role of ethics in architectural design can be treated differently from its historical position. Rather than justification, metaphor, or analogy, it becomes one of the many factors in the decision-making process of a multifactorial problem. The outcome of architectural design problems is a compromise involving choices between incompatible ends, as well as being limited by contingencies and coincidence. As such, it becomes relevant for architects to receive some basic training in ethical considerations, as some of their decisions may involve normative choices or value judgments. The typical approach to these normative choices in architecture has been to simplify the multiple relationships between value judgments, limited agency, and extrinsic contingencies. This simplification is one of the key features of twentieth-century architecture. Contemporary architecture discourse is in need of a more developed sense of ethical considerations by incorporating them both in the education and the research of architecture. Understanding ethics on

L. Schrijver (✉)
Faculty of Architecture, Delft University of Technology,
Delft, The Netherlands
e-mail: l.s.schrijver@tudelft.nl

C. Basta and S. Moroni (eds.), *Ethics, Design and Planning of the Built Environment*,
Urban and Landscape Perspectives 12, DOI 10.1007/978-94-007-5246-7_5,

a general level and arguing specific normative choices throughout both research and education in architecture will enable a more accurate understanding of the many factors in design decisions, including those that cannot be directly extrapolated from a program brief. In the domain of architecture, we may thus ask: which ethical considerations rightfully belong to the work of the architecture, where in the process of designing and building do these considerations come into play, and how may we understand and approach them?

5.2 Architecture: A Science of the Artificial

The practice of architecture is a complex activity, dependent on constraints within and without, and juggling multiple inputs (Till 2009). Reevaluating architecture as a valid field of academic research as opposed to a "minor profession" requires both a focus on the internal logic and vocabulary of the discipline, and an understanding of fields directly related or incorporated to the issues at hand. Architecture and the built environment form a complex whole, consisting of many different elements serving various functions. This precludes the possibility of finding a single optimal solution. It includes many parameters such as ethical, aesthetic, and sociopolitical concerns, which cannot be equated but rather need to be weighed in different manners. The problem of architectural design thus typically revolves around normative choices and a response to contradictory demands.

The status of architecture has been contested over the course of the twentieth century, whether it merits the term scientific research or is rather the result of creative inspiration. How architectural practice is viewed, and what the role of the architect is in relation to the built work and to society, circumscribes the scope of ethics in architecture. Nathan Glazer has coined the term "minor professions" to denote those professions that are non-rigorous, nonacademic disciplines, with an epistemological grounding that shifts, based on the contingencies of practice (Glazer 1974). Donald Schön, referring to the limitations of this model, introduced the idea of the "reflective practitioner," who combines intuitive practice and manual skill with reflection (Schön 1983). More recently, Nigel Cross has suggested we explore the "designerly way of knowing" as a third field alongside scientific knowledge and the knowledge of the humanities (Cross 2006). While science is concerned with the natural world, the humanities focus on human experience, and design then is oriented on the artificial world. To Cross, this rethinking offers a crucial reconfiguration of education, since those in what Glazer calls the "minor professions" are essentially trained but not educated: "activity which is cognitively adrift" (Cross 2006, p. 2). Referring to Ryle's distinction between "knowing how" and "knowing that," Cross sees design-related knowledge, rather than skill, as central to a reconsideration of design as a science. Earlier, Simon had already placed design at the center of human understanding: "The proper study of mankind is the science of design, not only as the professional component of a technical education, but as a core discipline for every liberally educated person" (Simon 1969, p. 159).

If architecture is to be considered a minor profession in the sense of Glazer, then it has a limited scope. The architect's responsibility is constrained to the object at hand, involving primarily proper conduct toward a client, and "diligence in performance" as Fil Hearn summarizes Ruskin's approach (Hearn 2003, p. 36). There is little appeal in this to standards of aesthetics, or to innovation, as those are external concerns unrelated to diligence. The sense of moral responsibility does not extend beyond the design and building process itself. In contrast, treating architecture as a "science of the artificial" (Simon 1969) expands its realm of responsibility significantly. The sciences of the artificial put forward alternative future scenarios, presupposing them to be desirable for any number of reasons. There are many factors involved, often in direct conflict with one another, each of which may be an object of research. The responsibility of the architect here is broader in scope, as it includes an envisioned result, which itself will influence its surroundings. This impact immediately implies ethical considerations being brought to the forefront of the design process, being incorporated in the various envisioned alternatives.

As a whole, architecture and the built environment pose a "larger" problem than many single-use objects: they are composed of many elements, serve many functions, and embody a level of complexity that precludes the possibility of finding a single optimal solution. Where "pure engineering" problems (small scale) may on occasion find a "best fit" solution, architecture and the city incorporate too many variables and (contradictory) factors to optimize the design solution. These factors are often not of equal standing in their relation to the design. There are external constraints to architecture, such as building regulations, available budget, and the various agents involved in the building process. There are also internal presuppositions, such as sociopolitical or sociocultural considerations (in the case of public buildings such as municipal offices or museums). In addition, there are the internal, design-based suppositions such as aesthetics or spatial arrangements. All of these elements are involved at different moments within the design process. This results in a complex discipline that shares characteristics with many other fields, yet cannot be reduced to these fields. The resulting confusion as to the status of architecture as a discipline – practice or discourse, engineering or art, minor profession or reflective practice – has resulted in a similar confusion of methodologies and approaches to research. An additional complication is the conception of the role of the architect. This might be as an artist, or the main "author" of a building, which preferences the ideas and the creative faculties of the architect over his role in the building process. This might be as a craftsman, treating the architect as a "master builder" who is extremely familiar with all aspects of construction. Or, as is increasingly common today, it might be as "process manager," in which the architect in the central figure in a field of complex processes such as communicating with the municipal authorities, the contractors, the client, and a project coordinator.

A central problem in the twentieth-century discourse on architecture is its faith in the analytic method of the natural sciences as a response to architectural problems. In the early twentieth century, this is visible in the references to engineering and the rational faculties of man. In the middle of the century, it is manifest in an increasing focus on the empirical findings of sociology (preferably quantified) and an interest in design methodology.[1] Yet both the natural sciences and the empirical findings of sociology have little to offer in terms of a normative guideline. A number of years ago, this was put forward by Bruno Latour in his distinction between "matters of fact" and "matters of concern," which suggested that the typical scientific view of modeled reality was insufficient to address the full breadth of human experience (Latour 2004). In a seminal form, however, Herbert Simon addresses this problem as the core of the "sciences of the artificial," by which he refers to domains such as engineering, computer science, and architecture. These domains are engaged with the artificial world, which they themselves contribute to. As such, they require the use of possible scenarios, necessitating a different approach than the purely analytic natural sciences. In the artificial sciences, one is always speculating on potential futures while simultaneously intervening in them. While Simon is focused mainly on what he calls "satisficing" (solving the problem in a satisfactory manner), the role of judgment is crucial in evaluating possible alternatives.

In this sense, designating architecture as a "science of the artificial" or understanding it to be based on a "designerly way of knowing" most closely aligns with architecture as a field of academic research without reducing it to analytic problem-solving. Following this view of architecture as a science of the artificial, a design problem is then not a matter of an optimal solution, but rather a question of choosing between fundamentally different possibilities. The guidelines for normative choices cannot be scientifically derived from spatial or programmatic analyses.[2] The design problem that is not objectively solvable requires normative choices in the decision-making process in order to choose between alternatives that resist direct or quantifiable comparison.[3] Thus, instead of implicit assumptions in the domain of ethics or a reduction to other areas such as program or style, it is necessary to treat ethical considerations as an autonomous contributing domain in the design process. Making these normative choices explicit can help qualify the final preferences or decisions. If architecture is a (non-optimizable) problem with incompatible requirements, the choices between the acceptable alternatives will typically be normative, requiring other categories of judgment. There are values inherent in many choices throughout the design process. They must be addressed from the perspective of design and desire: which factors do we consider more or less important in the long list of design decisions? Does the architect prioritize the user or the client? Cultural convention or progressive aesthetics? Programmatic organization or iconographic significance? While all these choices have specific design implications, they are founded on normative positions, which cannot be addressed solely through a design approach. In addition, the manner in which these values, both personal and societal, become embodied in the architectural object is not clear-cut. At the same time, architects need to focus on their own expertise, which lies within the realm of the spatial and aesthetic values.

In order to achieve this, the architect requires a basic competence in various fields and a basic grasp of the ethical issues that might factor in. In this sense, the expertise of the architect can be seen as a practice of synthesis, combining different areas of knowledge and bringing them into the design. Cross puts forward the strategy of synthesis as one of the fundamental traits of design practice. He refers to a 1979 experiment on design behavior that showed scientists as following a problem-focused, analytic strategy, while designers focused on the desired result (Cross 2006, p. 6). This entails a process of evaluation of alternatives within the process of problem-solving, without necessarily knowing all the variables. Finally it is important to realize that the architect is not omnipotent. Throughout the design process, the architect is required to incorporate different types of information into a cohesive building. These may include conditions such as building regulations, budget concerns, material characteristics, and the psychology of the user, to name but a few. Even if we agree with Alberti's characterization of the architect as a person of exceptional quality, capable of navigating all these domains of knowledge, there are external limits to the architect's influence. Placing full responsibility for a building with the architect presumes that the architect is in complete control and has the highest decision-making capabilities throughout the design and building process, which is often not the case. There may be a project manager involved to control the budget or to make final design decisions, or political regulations may present insurmountable obstacles for a design proposition. A characterization of architecture by Rem Koolhaas is perhaps more adequate to architecture, as a "hazardous mix of impotence and omnipotence" (Koolhaas 1995).

5.3 Ethics in Architecture, the Problem of Reduction

Over the course of the twentieth century, the role of ethical considerations has often been one of reduction or simplification. Ethical propositions have been used to justify aesthetic decisions, or to convince the client of a design proposition. The historical trajectory has often followed a path of extremes, positioning the architect either as the savior in societal dilemmas or as a target for societal complaints. In both cases, the relationship between ethical and moral assumptions and their later role in the resulting building has often been neatly sidestepped by focusing on general statements or by suggesting ethical qualities as metaphor. These statements were informed more by personal ideas on desirable behavior than by extensive philosophical research on the relationship between architecture and ethics. "If architecture is seen as something that can be truthful it must be immoral for it to tell a lie, and this belief runs through the French rationalists and the English Arts and Crafts theorists to twentieth-century propagandists.... The idea that what distinguishes one object from another is not style but morality has been very clearly stated by Pevsner who argues that 'sham materials and sham technique' are 'immoral'" (Watkin 1977, p. 4). Throughout many architecture treatises, there are a striking number of adjectives implying moral standing ("lofty goals," "sound judgment," "noble architecture"), as if this will naturally lead to an uncontestable style of architecture. Although this metaphoric mode of thinking is deeply embedded in architecture thinking, its implicit nature has also allowed it to be uncontested. While the avant-garde architecture of the early twentieth century was positioned as a moral obligation by its proponents, the rationalized destruction of the Second World War caused a turnaround. In an equally reductive gesture, the industrial language of modern architecture was seen to negate uniquely human qualities. Thus, the mid-century critics of early modernism turned to everyday society and nostalgia as representations of "good" architecture.[4] Here, the noble savage disparaged by Loos was put forward as the better human because it is uncontaminated by societal hypocrisy.

Striking in each of these positions is the direct analogy drawn between moral fiber, ethical positions, and architectural expression. This is a fundamental problem in architecture, which was identified in 1962 by sociologist Herbert Gans (1962) as "the fallacy of physical determinism." No matter what the architectural form proposed, each justification of it was founded upon implicit normative statements that were to find their ultimate expression in precisely this style and composition of architecture. By leaving these terms implicit, moreover, an accurate discussion of ethical considerations and their relation to the built environment were circumnavigated.

Current architectural criticism has drawn out a spectrum that places an ethical responsibility with architecture that denies its long-term cultural embedding and the autonomy of architectural form. The conflating of morality and aesthetics over the nineteenth and twentieth centuries indicates an underlying problem in architectural thinking: the loss of a formal and theoretical vocabulary with which to address the qualities of architecture more precisely and distinguish the embedding of moral statements from an aesthetic effect. Distinctions in these realms are vital, because when conflated they easily lead to the notion that, for example, beauty itself can form a moral guide, or that moral fiber automatically must lead to beautiful art. In the case of the socially engaged work of the 1960s, there was a notion that sociopolitical intent would automatically lead to good architecture. There are many parallel tendencies within architecture along the domains of aesthetics, ethics, and the sociopolitical fabric, each of which implicitly connects nonaesthetic values to aesthetic observations. These connections may not be necessarily wrong, but the implication that a certain proportion has an inherent ethical or social quality misreads the different domains within which we evaluate architecture.

In periods of primarily aesthetic focus, the foremost critique revolves around the lack of societal or cultural understanding. In the aftermath of postmodernism in particular, the focus on aesthetics has often been interpreted as a lack of concern for the societal embedding of architecture. For example, the title of the 2000 Biennale in Venice, "Less Aesthetics, More Ethics," immediately implies a trade-off along this spectrum. Following along the lines of Ruskin's suggestion that beauty follows from truthfulness, there is an underlying implication that ethical architecture will appear less aestheticized – and therefore perhaps also more beautiful? The notion in the Biennale is more directly heir to the 1960s notions that vernacular or unpolished architecture is more "true to nature" and therefore more beautiful in an authentic fashion, rather than catering to the whims of aesthetic dandyism. When aesthetic principles are seen to convey a moral position, the inverse is argued as well: the fluidity of aesthetic choices implies the instability of moral or ethical standards. It is in the wake of these ideological struggles in architecture that statements of value have become suspect.

Susan Sontag argues against this direct correlation between aesthetic and ethical concerns. "For the problem of art versus morality is a pseudo-problem. The distinction itself is a trap; its continued plausibility rests on not putting the ethical into question, but only the aesthetic. To argue on these grounds at all, seeking to defend the autonomy of the aesthetic (and I have, rather uneasily, done so myself), is already to grant something that should not be granted – namely, that there exist two independent sorts of response, the aesthetic and the ethical, which vie for our loyalty when we experience a work of art. As if during the experience one really had to choose between responsible and humane conduct, on the one hand, and the pleasurable stimulation of consciousness, on the other!" (Sontag 1965, p. 23). She acknowledges ethical and aesthetic concerns as related, but primarily on the level of our reception, both intellectual and sensual: "Art is connected with morality, I should argue .... The moral pleasure in art, as well as the moral service that art performs, consists in the intelligent gratification of consciousness" (Sontag 1965, p. 24).

In his 2004 article "Ethics versus Aesthetics in Architecture", Maurice Lagueux offers a fruitful insight along similar lines. The historical equivocation of ethics and aesthetics does not withstand close scrutiny, as it usually involves moral invocations as justification of aesthetic preferences. Lagueux notes that while ethical and aesthetic decisions are both internal to architecture, they cannot be characterized in identical terms. His tracing of the historical "fusion of beauty and morality" through the work of various architects and thinkers, such as Pugin, Ruskin, and Loos but also David Watkin, demonstrates that this coupling of aesthetics and ethics may typically serve any preconceived purpose. It is difficult to see ethical terms as objectively and causally linked to aesthetic intentions when they may serve such diverse preferences as the traditionalist architecture of Quinlan Terry and the modernist work of Marcel Breuer under ostensibly the same goals. In short, following Lagueux, "The paradoxical aspect of these *ethical* debates, is that they all bear on beauty, truth, and historical time much more than on good or on morality" (Lagueux 2004, p. 130).

An overinflated sense of societal value in twentieth-century architecture suggested that "appropriate" architecture would somehow magically resolve any variety of societal issues. In response the proposition of "autonomous architecture" arose, which essentially negated all ethical concerns. This compensates a flawed argument by arguing its inverse, that there are no ethical concerns relevant to architecture. Although historically it is understandable that a responsibility for the built environment is disavowed in the late twentieth century, it is unfortunate. The all-encompassing positions of (received) modernism resulted in a postmodernism that was resistant to sweeping statements, and reveled in specific aberrations. In its most extreme form, this results in the inverse problem: an inability to make any sensible statement that extends beyond the contingent, personal experience. This too seems to belie the history of shared human experience. It may thus be more sensible to speak of the limitations of architecture while acknowledging a need for responsible practice. There are domains of responsibility that do rest with the architect in his profession, while there are also (sociopolitical) domains that may be influenced by the built environment, but nevertheless cannot be resolved by it.

To recall the Koolhaas statement on architecture as both impotent and omnipotent, just because architecture has not fulfilled its early modern promise of moral guide does not mean that a building has no influence. It would be too easy to hide behind the notion that a building is "silent" and that a pile of bricks cannot communicate and therefore it has no morals. Buildings are not only instrumental, they are also tangibly present in our environment – they can influence how we perceive our environment or how we feel, we may get attached to them or dread our stay in them. The study of ethics appeals to a notion of the good beyond our immediate concerns. Incorporating this in the education of architects and research on architecture may contribute to a sense of the public good beyond the immediate requirements of the program.

If we wish to consider more carefully the role ethics might play in architecture, we should probably distinguish a few domains in the process of architectural design in which they have a place. First, there is the question of professional conduct. This is present in historical treatises as early as Vitruvius. It usually amounts to the reduction to specific issues such as professional conduct or the metaphorical representation of virtue. In *De re aedificatoria*, Leon Battista Alberti already positions the architect as a man of exceptional quality: "[The Architect] ought to be a Man of a fine Genius, of a great Application, of the best Education, of thorough Experience, and especially of strong Sense and sound Judgement" (Alberti 1452).

A clear distinction is made here between the conduct of the architect (the "performance value" that is part of making, as Ronald Dworkin (2011) calls it) and the final product. One may thus conceive that a building may be realized in accordance with aesthetic standards by an architect who is nevertheless unethical. Second, there is the question of the ethical properties of the building itself. To what extent can we envision ethical properties being embodied in the actual bricks? Yet, classical architecture nevertheless assumed that buildings can suggest values to their users. The Gothic cathedral is meant to appeal to a sense of humility in the face of God and to reach toward the heavens. Would this strictly speaking influence behavior? This domain is perhaps the most troublesome, as it requires a certain analogical reasoning or anthropomorphizing of the building.

Whether a material building actually has moral properties remains a contested issue. Can a spatial configuration communicate a moral position? In the late nineteenth century, Ruskin fulminates against "false representation" in architecture such as imitation granite. He believes that its only effect "is to cast suspicion upon the true stones below, and upon every bit of granite afterwards encountered" (Ruskin 1880, p. 48). In the early twentieth century, Loos is clear about the moral decadence of ornament, while Le Corbusier suggests that the engineer is to be admired for his clinical, rational products (Loos 1908; Le Corbusier 1927).

Third, there are the general and specific ethical considerations within the design process. These have no bearing on the professional conduct of the architect personally, but rather appeal to a sense of moral responsibility to the (built) environment. Today, this is most vivid in our concerns for the natural environment and issues of sustainable development in the building process. This domain can well be supplemented by more specific consideration of historical and contemporary values.

5.4 Incorporating Ethical Considerations in Design

It seems that the primary task for contemporary architecture is to explicate the value statements that are typically implicit in design. These may concern the concrete questions of design issues or moral principles such as diligence and responsibility. In particular, the ideas of the sciences of the artificial and designerly ways of knowing offer a suggestion toward avoiding the typical simplifications. Enumerating the many constraints of design and explicating the many design premises will open up a field in which aesthetic considerations, functional values, ethical concerns as well as financial constraints and building codes may all be discussed, with some being given preference over others. Given the impact of the built environment on the everyday lives of people, it would seem that education and research on ethics is crucial to any education in architecture and the building sciences. As such, an initial introduction of both descriptive and normative ethics may help guide the domains that would seem most pertinent to the study of architecture, both as profession and as academic discipline. By systematically introducing aspects of descriptive ethics, based on both the historical and the contemporary, we could identify how moral values have historically been ascribed to buildings, what we find appropriate to the built environment, and what kind of moral beliefs are embedded within it. Normative ethics would require a discussion of values: how do we propose it is best to live, to address our environmental issues, to allow for security, sustainability, freedom, and various other qualities of life often seen as desirable. Combining these two as explicit aspects of the study of architecture may help introduce a higher precision in design decisions, not as a guidance in terms of what is "correct" (as that is impossible in this field), but a better understanding of underlying assumptions and potential consequences of design decisions.[5]

Descriptive statements delineate the many possibilities, while normative statements evaluate the possibilities and identify preferred scenarios. Normative statements cannot be derived from analysis. Rather, they are founded on many considerations from personal beliefs to communal concerns and cultural values. Ethical positions in architecture are often metaphorical or analogical.[6] Though metaphor and analogy are interesting in a historical study and in terms of a "culture of design," perhaps it is more relevant to study ethical positions as a design component, as properties ascribed to buildings, or as aspects of the design process. The descriptive study of values incorporates an assessment of what is typically held to be valuable within a certain context. Descriptive ethics in this sense would primarily concern historical study and precedent analysis. This might contrast the values describing buildings, such as the notions of "efficiency" and "rationality" underlying the CIAM planning principles. Normative statements show what we aspire to, often taking the form of what "should be done," incorporating a value judgment. This requires an explication of the values involved, whether they are founded on socio-political concerns or individual moral positions.[7]

In light of a more important role for ethics in architecture education and research, a few domains of ethical inquiry in architecture might be distinguished. First, there is the domain of professional conduct, which resides in a personal view of performance, but also is embedded in professional regulations such as codes of conduct. Which primary features have been treated as central to professional conduct? How far does the professional responsibility of the architect extend, only to his client or also to society at large?

This domain is perhaps the most obvious, as it concerns the concrete ramifications of conduct. From the initial suggestions of Vitruvius and Alberti on the moral fiber necessary to the practice of architecture, to the book *The Ethical Architect* (Spector 2001), the role of the architect is seen as a demanding one, always bearing the public good in mind even when fulfilling a client's particular desires. The historical shift from an internalized professional ethics to the external regulations of professional organizations and their rules of conduct has not necessarily diminished the sense of responsibility of practicing architects, although it may have complicated their practice. In fact, the sense of responsibility was not explicitly present in the twentieth-century manifestoes, but all the more tangible for the forcefulness of the architect as vanguard. The rules of conduct stipulated by professional organizations largely adhere to "diligence in performance," implying that diligence must lead to good performance. This in itself is a questionable assumption, as it does not reflect on the ends of the project. Spector too points out the conflicts within societal requirements, noting that some issues cannot be solved through good design. He thus notes that some problems may not reflect on the architect's skills or, indeed, even his diligence (Spector 2001, p. 21).

The ethical properties of the building itself present more difficulties. This first separates into the ethical properties ascribed by architects and those ascribed by critics and the public. Architects, for example, often refer to the Vitruvian properties *utilitas*, *firmitas*, and *venustas*, as timeless properties of the art of construction.[8] This raises the following question: to what extent are these traits considered moral? They indicate what is seen to be of value in architecture, but they are also quite general, saying little about the design as such. In this, they offer a starting point, delineating what might be considered unacceptable (such as a building that collapses). It is, however, necessary to study examples to refine these general notions (*firmitas* might then be recast as structurally sound buildings with delicate proportions, such as Gothic cathedrals or the Eiffel tower, or expressively solid constructions such as the Pantheon or rather feats of structural bravado such as the Nationalgalerie's floating roof).

In this sense, case studies offer a palette of ethical properties historically ascribed to buildings. This approach risks again conflating aesthetic and ethical properties, or using ethical arguments in support of aesthetic preferences. Alain de Botton resolves this issue by identifying "aesthetic virtues" in analogy to personal virtues: "Analogising architecture with ethics helps us to discern that there is unlikely ever to be a single source of beauty in a building, just as no one quality can ever underpin excellence in a person. Traits need to arise at congruous moments, and in particular combinations, to be effective .... Armed with a comprehensive list of aesthetic virtues, architects and their clients would be freed from overreliance on Romantic myths concerning the chance or divine origins of beauty. With virtues better defined and more readily integrated into architectural discussions, we would stand a fairer chance of systematically understanding and re-creating the environments we intuitively love" (De Botton 2007, p. 174).

Here again, we find the ethical standard as an analogy to the aesthetic standard. The relationship between the two domains cannot be dismissed, yet may also not be oversimplified. One might consider such features as a "welcoming" building in the case of a public institution, or a "protective" building in the case of a housing complex or a single dwelling. "Daring" or "innovative" might work as triggers to our own imagination. Yet again these are formulated in general terms and require the specificity of case studies to refine them. While a blind wall that hides an entrance may not be a feature typically associated with welcoming, there may be instances in which this is designed as a protective entrance.

An important dilemma in this field is also the question of the building as "agent." How might we consider buildings as encouraging ethical behavior? The tendencies toward physical determinism of the members of Team X and Le Corbusier, for example, suggest that buildings, when properly designed, can induce desired behavior. In contrast, Sontag notes: "A work of art, so as it is a work of art, cannot – whatever the artist's personal intentions – advocate anything at all" (Sontag 1965, p. 26). One argument that has become popular recently is the idea of knowledge residing within objects. The most radical version of this is perhaps Bruno Latour's actor-network theory. Yet Nigel Cross also argues that there is knowledge residing within objects; he refers the Pye's notion that making or doing always precedes theory (Cross 2006, p. 9). If we assume this to be so, then the "agency" of buildings might be the knowledge that is communicated. This knowledge is laden with an implicit value: if the object has been created, it must be knowledge worth having. Here again we encounter the problem that knowledge and norms are not necessarily the same, nor do intent and design decisions necessarily translate into "agents" of ethical behavior.

In terms of the speculative assumptions of the discipline of design, we could study the assumptions put forward in a design and evaluate the results in retrospect. For example, the Pruitt-Igoe housing complex in St Louis Missouri was built in accordance with modernist principles of design. An underlying supposition in this type of design was that it appealed to the rational behavior of its inhabitants (as the "highest" faculties of human beings, according to Le Corbusier). A significant portion of the complex was demolished in 1972 after years of social turmoil. The inhabitants set fires in the hallways and urinated in the elevators, hardly a shining example of rational behavior.[9] Evaluating historical examples as such will remain specific instances, yet it might suggest general rules by induction.[10] In short, the ethical properties of a building must appeal to an idea of "the good," but how that is given concrete form may differ greatly from one instance to another. As such, the appeal to a notion of virtue, or the good, beyond immediate concerns, trivialities, or self-interest, is intuitively desirable and almost always an underlying intention in architecture. The architectural project, however, once constructed, often has unforeseen consequences that affect how a building is used, appreciated, remembered and valued. For this reason, a more specific discussion on ethical standards, in terms of the professional conduct, design premises, and the resulting built form, is required.

Notes

1. For example, Alison and Peter Smithson worked with the findings of sociologists Young and Wilmott in the 1950s. Design methodologies were central to many of the North American schools of architecture, MIT in particular.
2. The same thing can be said about aesthetic choices. The modernist argument on the "spirit of the age" notwithstanding, one would be hard-pressed to find a definitive causal connection between scientific analysis of design problems and the aesthetic proposition in a design solution.
3. Interestingly, the process for European project bids attempts to quantify various incommensurable aspects of architectural design by attributing a numerical value scale to such diverse characteristics as size of the organization, experience with a type of building, and design quality. Recently, the architecture office Kempe Thill has done research on this form of project bids and a proposition for improving them. The study is available online at http://www.atelierkempethill.com/0077_nl.pdf
4. The mid-century critique of dogmatic modernism demonstrated some crucial problems, but was equally reductive in its principles of sociability in architecture (Schrijver 2009, pp. 85–94).
5. One might also distinguish between the meta-discourse of ethics as relevant to architecture discourse and education, and applied ethics as relevant to specific designs or case studies, following the distinction by Fox (2000).
6. Nigel Cross has argued that metaphor and analogy are approaches that properly belong to the humanities, which concern human experience. In his classification of design science, it is pattern seeking and modeling that are essentially design approaches.
7. These two approaches together may help study contemporary developments, such as the increasingly economic interpretation of value. The current assumption that if something sells, it must be valuable uses economic value as an indicator of general value. For example, in a discussion of a housing project on Ypenburg by Christian Rapp, the criticism was directed primarily at the fact that over a fifth of the apartments were not yet sold. "Duur, saai, en niet eens een balkon" ("Expensive, boring, and not even a balcony") *NRC Handelsblad*, 17 November 2007. Assessing examples like this for their underlying assumptions on value may help reconsider current practices, such as the intervention of project managers who make decisions in the design process based primarily on economic considerations.
8. Spector, for example, uses the Vitruvian terms as his basis and expands them with "context" and "site" (Spector 2001).
9. This was famously put forward by Charles Jencks as the definitive end of modernism in architecture (Jencks 1977, p. 9).
10. For example, there is a general tendency toward physical determinism to be found in the history of twentieth-century architecture, particularly in the manifestoes and position statements.

References

Alberti LB (1987) De re aedificatoria. Dover Publications, New York [orig. 1452]
Cross N (2006) Designerly ways of knowing. Springer, London
de Botton A (2007) The architecture of happiness. Penguin, London
Dworkin R (2011) What is a good life? New York Review of Books, 58(2) February 10
Fox W (ed) (2000) Ethics and the built environment. Routledge, London
Gans H (1962) Urban vitality and the fallacy of physical determinism. In: People and plans. Basic Books, New York
Glazer N (1974) The schools of the minor professions. Minerva 12(3):346–364

Hearn F (2003) Ideas that shaped buildings. MIT Press, Cambridge, MA
Jencks C (1977) The language of post-modern architecture. Academy Editions, London
Koolhaas R (1995) SMLXL. 010 Publishers, Rotterdam
Lagueux M (2004) Ethics versus aesthetics in architecture. Philos Forum 2:117–133
Latour B (2004) Why has critique run out of steam? Crit Inq 30(2):225–248
Le Corbusier (1986) Towards a new architecture. Dover Publications, New York [orig: *Vers Une Architecture*, 1927]
Loos A (1908) Ornament and crime. In: Conrads U (ed) Programs and manifestoes of twentieth century architecture. MIT Press, Cambridge, MA, pp 19–24, 1970 [orig Frankfurt, Verlag Ullstein, 1964]
Ruskin J (1880) The seven lamps of architecture. George Allen, Kent [orig. 1849]
Schön DA (1983) The reflective practitioner: how professionals think in action. Basic Books, New York
Schrijver L (2009) Radical games: popping the bubble of 1960s Architecture. NAi Publishers, Rotterdam
Simon HA (1969) The sciences of the artificial. MIT Press, Cambridge, MA
Sontag S (1965) On style. In: Against interpretation and other essays. Picador, New York
Spector T (2001) The ethical architect. Princeton Architectural Press, New York
Till J (2009) Architecture depends. MIT Press, Cambridge, MA
Watkin D (1977) Morality and architecture. Oxford University Press, Oxford

Chapter 6
Aesthetics as a Risk Factor in Designing Architecture

Sabine Roeser

6.1 Introduction

This chapter argues that in designing architecture, aesthetics should be considered as a "risk factor" in addition to conventional risk factors such as environment and health and ethical considerations such as justice, fairness and equity. Buildings should not only be designed to be sustainable in a technical sense but also in an aesthetic sense.

Sustainability is concerned with minimizing risks of technologies for future generations. It has become an important criterion in designing architecture. Architects try to design buildings that are energy efficient, by, for example, using insulating materials and solar panels. They try to use sustainable and environmentally friendly materials and materials that do not create risks for the health of the users and builders of buildings. These are ways in which architects try to minimize the risks that buildings can pose for humans and the environment. By so doing, architects take on their moral responsibility to contribute to a sustainable and safe world.

However, this chapter argues that buildings should not only be designed to be sustainable in a technical sense, by avoiding pollution and health risks, but also in an aesthetic sense. Architecture differs from many other technical artefacts because

Work for this chapter has been done during a NIAS-Fellowship and with funding by NWO (the Netherlands Organization for Scientific Research) under grant number 276-20-012. I wish to thank the audience at a NIAS seminar and at a workshop on ethics of the built environment at TU Delft for their helpful comments on earlier versions of this chapter. I also wish to thank Lara Schrijver for drawing my attention on some of the references of this contribution.

S. Roeser (✉)
Faculty of Technology, Policy and Management, Section Philosophy/3TU. Centre for Ethics and Technology, Delft University of Technology, Delft, The Netherlands
e-mail: S.Roeser@tudelft.nl

C. Basta and S. Moroni (eds.), *Ethics, Design and Planning of the Built Environment*, Urban and Landscape Perspectives 12, DOI 10.1007/978-94-007-5246-7_6,

of its unavoidable presence in the (urban) landscape. Where in the case of most technological artefacts, the main concern is that they do not create unacceptable risks for health and environment, in the case of buildings, a concern should also be that they do not risk to be aesthetically outdated in a few years. The aesthetic risk of architecture is to be an eyesore in the landscape. This is where aesthetics and ethics directly intersect: an environment that is spoiled by aesthetically non-sustainable architecture affects our well-being. This contribution will argue that aesthetics should be included as a factor in thinking about risks and sustainability, especially in the context of architecture. Architects should strive to design "aesthetically sustainable architecture", in addition to designing environmentally sustainable architecture in the more conventional sense.

The challenge for architects is to be aesthetically innovative and yet to anticipate a timeless aesthetic taste. This chapter will propose that architects should take into account the concerns of the public in their designs, and that they should use their aesthetic and moral emotions in order to meet the challenge of designing aesthetically sustainable yet innovative architecture.

6.2 Qualitative Risk Factors

This section starts with comparing technocratic and alternative approaches to risk, where the former focus on quantitative aspects of risk such as annual fatalities and the latter focus on qualitative, for example, ethical aspects. In the next section it will be argued that in the case of architecture, the qualitative approach has to be supplemented with aesthetic values in addition to ethical values.

In conventional approaches to risk management, risk is defined as the probability of an unwanted effect. An example is the probability of annual fatalities as a consequence of a technology, for example, through accidents or pollution. Policy makers then apply cost/benefit analysis in order to determine whether a technology should be implemented. Proponents of such an approach praise it as a rational, objective and value-neutral method. However, this view has come under pressure.

During the last couple of decades, a lot of research has been done into qualitative and ethical aspects of risk. This research has been initiated by psychologists and social scientists, but recently, more and more philosophers have become interested in this topic. What these scholars have in common is a critical stance towards technocratic methods of risk assessment that are based on, for example, cost-benefit analysis.

Empirical research has shown that laypeople have a broader conception of risk than experts (Fischhoff et al. 1981). They include qualitative, ethical considerations in their intuitive conception of risk, such as justice, fairness, autonomy and whether a risk is catastrophic or not (Slovic 2000). Philosophers emphasize that these are important ethical considerations in thinking about risks (Hansson 2004). A richer understanding of the notion "risk" is needed. The moral considerations of the public are reasonable and should be taken into account in ethical decision-making about risky technologies (Kahan and Slovic 2006). These ideas have resulted in various approaches to participatory technology assessment that aim to involve the public in decision-making about risky technologies, such as town hall meetings, roundtable discussions and scenario development (Sclove 1996; Van Asselt and Rijkens-Klomp 2002; Gregory and Keeney 1994; Boenink et al. 2010).

6.3 Aesthetics as a "Qualitative Risk Factor"

When it comes to risky technologies, architecture is a special case. Most technical artefacts can create risks for health and the environment. However, in the case of architecture, there is an additional dimension involved, that is, an *aesthetic* dimension. Architecture creates artefacts that influence our *visual* environment (Nasar 1988b; Carlson 2000; Scruton 1979; Hill 1999). Some buildings are an eyesore in the (urban) landscape. Venturi et al. use the notion of "visual pollution" for this phenomenon (Venturi et al. 1972). In contrast, other buildings are beautiful and contribute positively to the aesthetic quality of a city or town. Buildings matter aesthetically. But this fact also influences the quality of life and well-being in an area. Here, ethical and aesthetic concerns coincide (Harries 1997; Taylor 2000). Hence, buildings should be sustainable not only from the point of view of health and environment but also from an *aesthetic* point of view.

The importance of aesthetics for well-being is widely acknowledged in architecture theory, and many scholars and architects are aware of the moral responsibility that architects have in order to design buildings that make a positive aesthetic contribution to society. However, in the literature on technological risk, aesthetics has not yet been acknowledged as a factor that poses threats or risks to people's well-being. In other words, aesthetics has not yet been conceptualized as a risk factor. However, such a conceptualization can lead to fruitful insights. This chapter aims to connect the discussion on the moral responsibility of architects to design aesthetically acceptable buildings with the discussion in risk theory on which qualitative risk factors to include in the conception of morally acceptable risks. The proposal made in this section is to include aesthetics as a qualitative risk factor in the discourse on risk ethics, as it makes an important addition to the existing lists of qualitative risk factors.

This is how aesthetics fits within the risk-ethics framework: (i) as stated above, bad aesthetics can affect our well-being, which is where ethics and aesthetics intersect. (ii) Aesthetics comprises uncertainty: How will people appreciate the aesthetics of a building a few decades from now? (iii) Bad aesthetics threatens future generations by burdening them with potentially horrendous buildings. (iv) Aesthetic risks give rise to issues of justice, fairness and autonomy: a wealthy person has more freedom and power to choose in which aesthetic environment he/she wants to live, just as he/she has more freedom and power to choose a healthy and sustainable environment to live in, than a poor person. Inasmuch as aesthetics contributes to people's quality of life and well-being, injustices concerning other aspects of quality of life and well-being also play a role in the context of aesthetics.

Authors who write on risk management, risk perception and risk ethics have largely neglected the aesthetic dimension of technologies (exceptions who mention aesthetics as a determinant in risk perception are Willis et al. 2005 and Willis and DeKay 2007). This is a major omission. There is a research community on environmental aesthetics (Carlson 2007, in Zalta 2007), but that discourse exists separately from the risk-ethics discourse. It would be fruitful to bring these two discussions together. The two discourses can profit from their respective insights and expertise. Specifically, it will enable risk scholars to have a more complete account of ethically relevant considerations in thinking about risky technologies.

The conventional, technocratic approach defines risk as a probability of an unwanted effect and then applies cost-benefit analysis to determine which risky activity has the lowest net risk. Ethical considerations such as justice, fairness and autonomy put boundaries on the cost-benefit analysis, mirroring the deontological and virtue ethical objections against consequentialist approaches in ethics. But ethical considerations also play a role in the determination of which kinds of effects to take into account in a risk assessment, for example, concerning the questions whether only to look at annual fatalities or also at sick or injured people, or at effects to nature. The proposal to take into account aesthetic considerations in risk assessment also concerns the kinds of effects to take into account. It broadens the scope of morally relevant consequences in risk assessment from matters of life, death, health and physical well-being to aesthetic well-being.

Connecting the two debates of aesthetic responsibility of architects and risk ethics is more than a merely theoretical exercise. As these two theoretical approaches have direct practical implications for professionals and policy makers, the connection of the two approaches will make their practical relevance even more explicit. More specifically, by leaving aesthetics out of the current debate about acceptable risk, there is a danger that aesthetics will be left out of the decision procedure. However, it is important to include aesthetic considerations next to other morally relevant considerations from the outset in order to proactively reflect on how to do justice to the various morally relevant risk considerations, including aesthetic considerations. By leaving them out of the "risk equation", there is the danger that qualitative risk aspects come in only as an afterthought. It is much more effective to take into account qualitative risk considerations such as justice, fairness, autonomy, equity and aesthetics from the start. This will enable to find the best solutions to doing justice to them all in the most optimal, morally responsible way.

Taking into account aesthetics in thinking about risky technologies allows considering a broader range of morally relevant features in risk assessment. These ideas can be extended to urban planning and to aesthetic aspects of nature that might be threatened by human activities. Here are a couple of examples. When urban planners have to design a highway, they have to take into account the amount of noise and emissions that might affect the health of people living in the vicinity, but they also have to take into account the visual effect that the highway has on the landscape. A highway is more controversial when it is planned to cut through a natural resort rather than through an area that is already built in a functional way. Another example concerns windmills and solar cells. Wind energy and solar energy are more sustainable than other sources of energy on the dimension of CO_2 emissions.

However, critics of these sources of energy point to the fact that massive parks of windmills or solar cells can harm the *visual* aspects of our environment. This is an important consideration that has to be given due weight. This does not imply to abandon windmills and solar cells because of their potential to harm the environment aesthetically; rather, in planning appropriate locations for these technologies, one should give this sufficient consideration. For example, rather than placing windmills in the middle of some rolling hills, one could place them along highways or in the area of a harbour, as these are already visibly human-made, industrialized environments. In the case of placing solar cells on the roofs of buildings, the solar cells should be designed such that they do not interfere with the design of the building. They should either be as invisible as possible or maybe even contribute positively to the design of a building.

These examples illustrate that aesthetic aspects already play an important role in debates about environmental issues. However, more conceptual clarity can be achieved by explicitly including aesthetic aspects in thinking about the ethics of risky technologies. Otherwise, there is a danger that because aesthetic considerations are not part of the conceptual framework for risk assessment, they will be left out as being supposedly irrelevant. The methodologies for assessing risky technologies have to be formulated in such a way that all morally relevant considerations are given due weight. This means that the ethics of risk has to be extended with aesthetics as yet another qualitative risk factor that we have to take explicitly into account.

This gives rise to the question how to balance the different ethical considerations in a risk assessment. Where apparently cost-benefit analysis provides us with a clear-cut, quantifiable methodology, alternative approaches to risk leave open how to trade off considerations of equity and fairness with considerations of overall, aggregate well-being. Some propose to design models that give specific weight to the various factors. However, this can seem like an ad hoc solution to a much more fundamental problem. It is unclear whether it can be determined in advance how to balance various potentially conflicting ethical considerations. There is a huge debate in meta-ethics on this question. Virtue ethicists, particularists and other defenders of "context-sensitive" approaches argue that it has to be judged on a case-by-case basis how to balance different ethical considerations (McDowell 1998; Dancy 2004). Various philosophers of risk argue that consequentialist approaches such as cost-benefit analysis run into serious methodological problems, next to the already mentioned ethical problems that relate to disregarding issues of distribution (Hansson 2004; Shrader-Frechette 1991; Asveld and Roeser 2009). The methodological problems concern the question how to measure and compare different sorts of well-being and how to value a human life. The methodological assumptions in technocratic approaches to risk are often highly arbitrary and can make a huge difference to the comparative assessment of various risky activities. For example, depending on how high one values a human life in a cost-benefit analysis can lead to diametrically opposed outcomes. Cost-benefit analysis gives us an illusion of objectivity, blurring the underlying substantial ethical considerations rather than making these explicit and subject of critical deliberation. Risk decisions are inherently morally complex. It is better to face this moral complexity explicitly, even though this means that one cannot fall back on a given, clear-cut methodology. Modelling moral trade-offs *requires* explicit moral reflection; it cannot *replace* it. Modelling moral trade-offs can then at most be a tool for ethical reflection in an iterative process.

In the case of including aesthetic considerations in a risk assessment, the fundamental question is how to trade off aesthetic considerations versus considerations of physical well-being of humans and nature. Although it is not theoretically impossible to design a quantitative model that makes such trade-offs, the philosophical question remains whether these are morally justified trade-offs and whether the same trade-offs can be made in each and every case. Because these are legitimate open questions, the proposal of this section is to bite the bullet and accept that broadening the scope of risk assessment of architecture to aesthetic considerations will inevitably involve contextual, situation-specific deliberation. Such a process of deliberation should involve different stakeholders, as they can provide for a broad range of ethical insights (Roeser 2007). The following sections will sketch the outlines of a framework of such a deliberation. This will involve a discussion of roles of the various actors that should be involved in this deliberation: architects, the public and policy makers (Sect. 6.4), and the capacities needed for this deliberation: moral and aesthetic emotions (Sect. 6.5).

6.4 Aesthetically Sustainable Architecture

How should architects deal with the proposed insight that aesthetics is a risk factor? The challenge for architects is to take into account a timeless aesthetic taste. They should strive to design buildings that are sustainable not only from an environmental or health point of view but also from an aesthetic point of view. They should design buildings that do not just live up to the latest fashion, but that also presumably will be appreciated by people who live in the future.

Now on a pessimistic view, one might think that it is impossible to predict how people will perceive buildings in the future. For example, according to postmodern or social-constructivist approaches to aesthetics, there is nothing like objective aesthetic criteria, aesthetics is merely a social construction, and it is hard to predict what people will appreciate in the future. Or one might be a sceptic: even though there might be objective aesthetic criteria, people might just not grasp them. On such views, aesthetics might be more a matter of "ignorance" rather than risk: in the risk literature, authors use the notion ignorance for future effects of technologies that are beyond prediction.

However, these are too pessimistic views. There are aesthetic theories that propose that aesthetic values are objective and that people can have access to these aesthetic values (for an objectivist account of environmental aesthetics, see Carlson 2000). This view is analogous to the theory of realism in meta-ethics, according to which moral values are objective and people can have knowledge of them (Shafer-Landau 2003; Cuneo 2007; Roeser 2011).

Note that also on an objectivist, realistic view of moral and aesthetic values, it is possible to err or to disagree. Just like in all forms of knowledge, aesthetic and moral knowledge is fallible. In addition, there is always room for reasonable disagreement on ethical issues, and so there is in aesthetics, as there can be a whole range of "aesthetically acceptable" solutions that allows for subjective preferences. That is exactly why it is impossible to predict with certainty what will be aesthetic ideals that will be shared universally across time and space. But just as there can be an overlapping consensus in ethics (Rawls 1996), so there can be one in aesthetics (note that various contributions in Nasar 1988 testify consensus in judgments about environmental aesthetics).

Some moral and aesthetic options are off limits. For example, in ethics, the objective boundaries are clear violations of human rights (Roeser 2006a; Rachels 1999; Wellman 1963; Moser and Carson 2001). In aesthetics, there are also clear cases where many people agree. Nasar provides for empirical evidence for this based on a comparative study between American and Japanese subjects and their aesthetic judgments of urban street scenes (Nasar 1988a, b). As Isaacs formulates it: "Each of us possesses an aesthetic instinct that is, at the most basic level, common to all, but moulded by individual and cultural experience" (Isaacs 2000).

Theorists of architecture and aesthetics have proposed various design principles that are supposed to contribute to the aesthetic quality of a building. Examples of such principles or features are a balance between order on the one hand and variation and disruption on the other (Gombrich 1984; Kaplan and Kaplan 1982; Berlyne 1974) or a balance between prospect (a wide view) and refuge (protection) (Appleton 1975). Nygaard Folkmann proposes the following two aspects of aesthetics in design: design as a structure of sensual appearance and design as an act of communication of an idea (Folkmann 2009).

A problem is that there is often a gap between the aesthetic assessment of experts and that of laypeople (Hershberger 1988; Gifford et al. 2000). In the case of an art work that is placed in a museum, this is not so problematic, as one can just avoid looking at the work in question. However, in the case of architecture or other aesthetic objects that are part of our lived environment, people are confronted with these objects every day. Hence, designers of such objects have a responsibility to take into account the aesthetic preferences of people who occupy these public spaces. This includes users of a building as much as spectators who regularly or incidentally pass by a certain building, as they are all more or less directly affected by the aesthetic aspects of a building. That is exactly the ethical dimension of aesthetics that I mentioned earlier. For this reason, architects cannot afford the same elitist aesthetic ideals as artists who create objects that people can avoid looking at (Lampugnani 2006). Architecture is not "l'art pour l'art", but it is an unavoidable part of our daily life and in that sense closer related to consumer design. Fortunately, there is research into aesthetic perceptions of laypeople which is meant to help architects to take into account the effect their designs have on laypeople (Hershberger and Cass 1988; Groat 1988).

This understanding of aesthetics enforces the idea proposed in this chapter that aesthetics fits well within risk discourse: there is uncertainty about universal aesthetic values, but not complete ignorance, albeit that it might be impossible to quantify or assign probabilities to aesthetic values (although see Carlson 1977). However, that holds for the other qualitative aspects of risk as well, and that is one of the ways in which they are distinguished from descriptive aspects of risk that can be measured and statistically monitored.

There are examples of architecture that seem to succeed in being aesthetically sustainable. Of course, there is always room for disagreement and exceptions, like on all matters, and maybe especially on evaluative matters. However, here are a few examples of architecture that are presumably rather uncontroversial. Think of the old downtown areas of many European cities, or of the temples in Tibet and other Asian countries or of the magnificent architecture of the Mayas and the Egyptian pyramids. These places attract millions of tourists year after year. Examples of modern buildings in, for example, the Netherlands that incorporate a timeless aesthetic ideal are buildings from the "Amsterdam school" from the early twentieth century and family homes from the 1930s that are still immensely popular.

So it seems possible to build in an "aesthetically sustainable way". At the same time, it is a challenge for architects to nevertheless be aesthetically innovative. A safe but not innovative way of building is to imitate a well-established and popular style. In the Netherlands, in new suburbs there are currently a lot of retro-1930s houses or areas that imitate the famous Amsterdam canals from the seventeenth century. However, this seems too much "the easy way out". Here is an analogy with the precautionary principle from mainstream risk discourse: retro-architecture chooses to rather be "safe than sorry". But a well-known possible pitfall of the precautionary principle is that it might preserve a status quo that could actually be improved by new technological developments. As Sunstein (2005) has pointed out, holding on to the status quo can be just as risky as innovation can be. This insight from mainstream risk discourse can be directly applied to aesthetic risk: avoiding aesthetic innovation and its concomitant risks can lead to aesthetic conservatism, which is also a risk, by foreclosing the possibility of better designs. This means that avoiding one aesthetic risk can lead to another aesthetic risk. In the case of aesthetic aspects of architecture, architects can add something to the experience of people by exploring new boundaries.

Similar ideas are discussed by Van de Poel in the wider context of design for well-being (Van de Poel 2012). He distinguishes the two main approaches to well-being and their relevance for engineering design. These approaches to well-being are desire satisfaction theories and objective list accounts. Van de Poel rejects the first account because people may desire things that are contrary to their own or other people's well-being. Instead, he defends an objective list account. Designers have a moral responsibility to design for morally defensible accounts of well-being. These ideas can be extended to architecture design: retro-architecture satisfies the desires of costumers, but it does not provide them with something that might in the end be even better (d'Anjou 2010).

This might invite the worry that such an approach might lead to a paternalistic attitude of architects. However, a normative approach to values in design can still incorporate the considerations of stakeholders, without taking these to be ultimately authoritative. For example, an explicit phase of reflection and deliberation can be included in design to take into account stakeholder values and also allow for normative reflection (Manders-Huits and Zimmer 2009).

The previously stated analogy with the precautionary principle can be helpful here. An important aspect of the precautionary principle is a reversal of the burden of proof. Rather than that opponents have to prove that a new technology is dangerous, proponents have to show that it is safe. In the case of the aesthetic risks of architecture, arguably architects have to do a good job to convince the public that their innovative designs are an improvement to more conventional designs that have proven their merit. This includes an educational task for architects, that is, to help people see the merits of new developments in architecture. Architects have to show the public that their designs are aesthetically sustainable. Here they can be aided by architecture critics who might be better capable in articulating the merits of a building than the architects themselves. If architects fail to convince the public, they have to come with something better.

This is where there is an important role for policy makers: they should design the institutional environment which allows for a dialogue between architects and the public, for example, by organizing town hall meetings and by providing information on new building projects in an accessible way. The policy makers have a special responsibility to enable a procedure that avoids the two pitfalls that threaten a genuine deliberation and dialogue about new technologies: one pitfall is that the experts are invoked as the ones who provide the ultimate answers; the other pitfall is that the will of the public is the ultimate arbiter. But in neither of these situations, there is a genuine dialogue and exchange of ideas. The former situation can be characterized as the "technocratic pitfall", the latter as the "populist pitfall". The technocratic pitfall sees science communication as a one-way approach. The populist pitfall considers majority vote the ultimum of democratic decision-making. However, in the so-called "deliberative democracy" approaches, the emphasis is on a genuine exchange of ideas, where the involved parties are open to each others' considerations and engage in what Habermas called a "machtsfreier Dialog" (Habermas 1996; Rawls 1996; Bohman and Rehg 1997; Guttman and Thompson 2000). In such a procedure, the outcomes are open, and a shift of positions is possible. Because all parties know that their concerns have received a fair hearing, they are more willing to reach a consensus that may involve compromises from all parties.

Architecture can be innovative and yet pleasing to a large audience at the same time; buildings can be designed to challenge our imagination and extend our ideas. In this way, architecture can diminish aesthetic risks and be "aesthetically sustainable", without being attached to a status quo that would inhibit one of the features that makes architecture important for people: by letting them explore new boundaries on how the buildings they live in and that they use can look like.

6.5 Aesthetic and Moral Emotions as Guides

The previous section suggested some procedures for coming to aesthetically sustainable architecture, such as taking into account certain aesthetic principles that are widely shared, including a reflection phase that takes into account concerns of stakeholders and shifting the burden of proof onto architects. In addition to these procedures, this section rounds off with the question by which epistemological means architects can best determine how a potentially "aesthetically sustainable building" should look like. The proposal is that in order to minimize aesthetic risks and design aesthetically sustainable buildings, architects should use moral and aesthetic emotions as guides.

Empirical studies have shown that emotions determine the risk perception of laypeople to a large degree (Alhakami and Slovic 1994; Finucane et al. 2000; Slovic et al. 2002, 2004). It seems to be a platitude that emotions are irrational, subjective and unreliable. This idea has led some scholars to scepticism towards the legitimacy of public risk perceptions. Most notably, Sunstein argues that since emotions are irrational, emotion-based considerations should be excluded from risk analysis and cost-benefit analysis should be used instead (Sunstein 2005).

However, on a different conception of emotions, this conclusion becomes dubious. According to various emotion scholars, both in psychology as in philosophy, emotions are necessary for practical rationality (Damasio 1994; Solomon 1993; Nussbaum 2001; Roberts 2003). This idea can be applied to emotions in risk perception. Emotions such as sympathy, fear, indignation and enthusiasm point to morally salient aspects of technologies, such as risks, benefits, autonomy and "fairness" (Roeser 2006b). Emotions help to make trade-offs between different ethical considerations, in general, but also in the case of risks (Roeser 2006b, 2009).

Emotions can play a role in designing for environmental sustainability. Chapman (2009) develops the notion of "emotional durability" for a design that leads to less wasteful behaviour by ensuring more stable relationships between consumers and products. However, emotions can also play a role in designing for aesthetic sustainability, as follows:

Empirical research indicates that architects have a hard time predicting the evaluations of laypeople of buildings (Brown and Gifford 2001). However, moral emotions can help architects to empathize and sympathize with potential users and to take into account their point of references and experiences (Greenbie 1988; Hershberger 1988; Roeser 2012). This requires that architects train their emotions and imagination so that they can put themselves into the shoes of a broad audience, also concerning future people.

In addition, architects should use *aesthetic* emotions. *Aesthetic* emotions can help architects to create sublime designs, designs that can bring people into a state of aesthetic delight. Several authors who write on environmental aesthetics in general and on the aesthetics of architecture specifically emphasize the importance of emotions in the aesthetic assessment of architecture (Gifford et al. 2000).

Nevertheless, in order to avoid the pitfall that architects get carried away by their own aesthetic emotions, they can use methodologies with which to predict the aesthetic emotions of users. Desmet et al. (2007) have developed a methodology to measure and predict emotions that industrial design products can elicit. This methodology could be extended to architecture.

Moral emotions can help architects to keep in mind the well-being of the users of their buildings and of people who will experience these buildings as part of their daily environment. Aesthetic and moral emotions can help architects to be sensitive to the aesthetic and moral needs of people. By giving emotions an important role, "aesthetic risks" can be minimized, and the way is open for an aesthetically sustainable architecture.

6.6 Conclusions

This chapter has argued that aesthetics should be conceptualized as a risk factor in designing architecture. The aesthetic risk of architecture is that it can become an eyesore that diminishes people's well-being. The moral significance of aesthetic aspects of architecture has been acknowledged by architects and architecture scholars for a long time. However, by conceptualizing aesthetics as a risk factor, important insights from the ethics of risk can be fruitfully applied to reflecting about morally responsible design of architecture. Aesthetics should be included in the list of qualitative risk factors that has been developed by philosophers and psychologists in recent years. By including aesthetics as a risk factor, it becomes more explicit that aesthetics has to be included in trade-offs of various risk aspects in designing architecture. The analogy with the precautionary principle in discussions about environmental sustainability can shed fruitful light on trade-offs between risks of innovation versus risks of conservatism in architecture. Trade-offs between various risk considerations require explicit ethical reflection that should involve a broad range of stakeholders. Ethical reflection about aesthetic risks can be facilitated by including moral and aesthetic emotions of designers and of stakeholders. This will enable the development of "aesthetically sustainable" architecture.

References

Alhakami AS, Slovic P (1994) A psychological study of the inverse relationship between perceived risk and perceived benefit. Risk Anal 14:1085–1096
Appleton J (1975) The experience of landscape. William Clowes, London
Asveld L, Roeser S (eds) (2009) The ethics of technological risk. Earthscan, London
Berlyne DE (1974) Studies in the new experimental aesthetics: steps toward an objective psychology of aesthetic appreciation. Halsted, New York
Boenink M, Swierstra T, Stemerding D (2010) Anticipating the interaction between technology and morality: a scenario study of experimenting with humans in bionanotechnology. Stud Ethics Law Technol 4(2):4. Available at: http://www.bepress.com/selt/vol4/iss2/art4
Bohman J, Rehg W (eds) (1997) Deliberative democracy: essays on reason and politics. MIT Press, Boston
Brown G, Gifford R (2001) Architects predict lay evaluations of large contemporary buildings: whose conceptual properties? J Environ Psychol 21:93–99
Carlson A (1977) On the possibility of quantifying scenic beauty. Landscape Plann 4:131–172
Carlson A (2000) Aesthetics and the environment: the appreciation of nature, art and architecture. Routledge, London
Carlson A (2007) Environmental aesthetics. In: Zalta EN (ed) The Stanford encyclopaedia of philosophy. SEP, Stanford. http://plato.stanford.edu/entries/environmental-aesthetics/
Chapman J (2009) Design for (emotional) durability. Des Issues 25:29–35
Cuneo T (2007) The normative web: an argument for moral realism. Oxford University Press, Oxford
d'Anjou P (2010) Beyond duty and virtue in design ethics. Des Issues 26:95–105
Damasio A (1994) Descartes' error. Putnam, New York
Dancy J (2004) Ethics without principles. Oxford University Press, Oxford

Desmet PMA, Porcelijn R, van Dijk M (2007) Emotional design; application of a research based design approach. J Knowl Technol Policy 20(3):141–155
Finucane M, Alhakami A, Slovic P, Johnson SM (2000) The affect heuristic in judgments of risks and benefits. J Behav Decis Mak 13:1–17
Fischhoff B, Lichtenstein S, Slovic P, Derby SL, Keeney R (1981) Acceptable risk. Cambridge University Press, Cambridge
Gifford R, Hine DH, Muller-Clemm W, Dárcy J, Reynolds JR, Shaw KT (2000) Decoding modern architecture: a lens model approach for understanding the aesthetic differences of architects and laypersons. Environ Behav 32:163–186
Gombrich EH (1984) The sense of order. Phaidon, London
Greenbie BB (1988) The landscape of social symbols. In: Nasar JL (ed) Environmental aesthetics: theory, research, and applications. Cambridge University Press, Cambridge, pp 64–73
Gregory R, Keeney RL (1994) Creating policy alternatives using stakeholder values. Manag Sci 8:1035–1048
Groat LN (1988) Contextual compatibility in architecture: an issue of personal taste? In: Nasar JL (ed) Environmental aesthetics: theory, research, and applications. Cambridge University Press, Cambridge, pp 228–253
Guttman A, Thompson D (2000) Democracy and disagreement. Harvard University Press, Cambridge
Habermas J (1996) Between facts and norms: contributions to a discourse theory of law and democracy. Blackwell, Oxford
Hansson SO (2004) Philosophical perspectives on risk. Techné 8:10–35
Harries K (1997) The ethical function of architecture. MIT Press, London
Hershberger RG (1988) A study of meaning and architecture. In: Nasar JL (ed) Environmental aesthetics: theory, research, and applications. Cambridge University Press, Cambridge, pp 175–194
Hershberger RG, Cass RC (1988) Predicting user responses to buildings. In: Nasar JL (ed) Environmental aesthetics: theory, research, and applications. Cambridge University Press, Cambridge, pp 195–211
Hill R (1999) Designs and their consequences: architecture and aesthetics. Yale University Press, New Haven/London
Isaacs R (2000) The urban picturesque: an aesthetic experience of urban pedestrian places. J Urban Des 5:145–180
Kahan DM, Slovic P (2006) Cultural evaluations of risk: "values" or "blunders"? Harv Law Rev 119:166–172
Kaplan S, Kaplan R (1982) Cognition and the environment: functioning in an uncertain world. Praeger, New York
Lampugnani V (2006) The city of tolerant normality. In: Graafland K (ed) Crossover. 010 publishers, Rotterdam, pp 294–312
Manders-Huits N, Zimmer M (2009) Values and pragmatic action: the challenges of introducing ethical intelligence in technical design communities. Int Rev Inform Ethics 10:37–44
McDowell J (1998) Mind, value, and reality. Harvard University Press, Harvard
Moser PK, Carson TL (eds) (2001) Moral relativism: a reader. Oxford University Press, New York/Oxford
Nasar JL (1988a) Visual preferences in urban street scenes: a cross-cultural comparison between Japan and the United States. In: Nasar JL (ed) Environmental aesthetics: theory, research, and applications. Cambridge University Press, Cambridge, pp 260–274
Nasar JL (ed) (1988b) Environmental aesthetics: theory, research, and applications. Cambridge University Press, Cambridge
Nussbaum M (2001) Upheavals of thought. Cambridge University Press, Cambridge
Nygaard Folkmann M (2009) Evaluating aesthetics in design: a phenomenological approach. Des Issues 26:40–53
Rachels J (1999) The challenge of cultural relativism. In: Rachels J (ed) The elements of moral philosophy, 3rd edn. Random House, New York
Rawls J (1996) Political liberalism. Columbia University Press, New York

Roberts RC (2003) Emotions. An essay in aid of moral psychology. Cambridge University Press, Cambridge
Roeser S (2006a) Intuitionism, moral truth, and tolerance. J Value Inq 39:75–87
Roeser S (2006b) The role of emotions in judging the moral acceptability of risks. Saf Sci 44:689–700
Roeser S (2007) Ethical intuitions about risks. Saf Sci Monit 11:1–30
Roeser S (2009) The relation between cognition and affect in moral judgments about risk. In: Asveld, Roeser (eds) The ethics of technological risk. Earthscan, London, pp 182–201
Roeser S (2011) Moral emotions and intuitions. Palgrave Macmillan, Basingstoke
Roeser S (2012) Emotional engineers: toward morally responsible engineering. Sci Eng Ethics 18(1):103–115
Sclove RE (1996) Town meetings on technology. Technol Rev 99:24–31
Scruton R (1979) The aesthetics of architecture. Mackay Limited, Chetham
Shafer-Landau R (2003) Moral realism: a defence. Oxford University Press, Oxford
Shrader-Frechette K (1991) Risk and rationality. University of California Press, Berkeley
Slovic P (2000) The perception of risk. Earthscan, London
Slovic P, Finucane M, Peters E, MacGregor DG (2002) The affect heuristic. In: Gilovich T, Griffin D, Kahnemann D (eds) Intuitive judgment: Heuristics and biases. Cambridge University Press, Cambridge, pp 397–420
Slovic P, Finucane M, Peters E, MacGregor DG (2004) Risk as analysis and risk as feelings: some thoughts about affect, reason, risk, and rationality. Risk Anal 24:311–322
Solomon R (1993) The passions: emotions and the meaning of life. Hackett, Indianapolis
Sunstein CR (2005) Laws of fear. Cambridge University Press, Cambridge
Taylor N (2000) Ethical arguments about the aesthetics of architecture. In: Fox W (ed) Ethics and the built environment. Routledge, London, pp 193–206
Van Asselt M, Rijkens-Klomp N (2002) A look in the mirror: reflection on participation in integrated assessment from a methodological perspective. Glob Environ Chang 12:167–184
van de Poel I (2012) Design and well-being. In: Brey P, Briggle A, Spence E (eds) Good life in a technological age. Routledge, New York
Venturi R, Brown DS, Izenour S (1972) Learning from Las Vegas. MIT Press, Cambridge, MA
Wellman C (1963) The ethical implications of cultural relativity. J Philos 60(7):169–184
Willis HH, DeKay ML (2007) The roles of group membership, beliefs, and norms in ecological risk perception. Risk Anal 27:1365–1380
Willis HH, DeKay ML, Fischhoff B, Granger M (2005) Aggregate, disaggregate, and hybrid analyses of ecological risk perceptions. Risk Anal 25:405–428

Chapter 7
Cost-Benefit Analysis and Evaluating Transport Safety Effects: A Discussion from the Perspective of Ethics

Bert van Wee and Piet Rietveld

7.1 Introduction

Transport is crucial for society: Societies cannot function without the transport of people and goods. It enables us to participate in many activities at different locations, such as living, working, education, shopping and visiting relatives and friends. In addition, it allows us to transport goods, from the locations of mining of raw materials, via several production stages, culminating in the shops where people buy products, or even up to the final locations of use, such as houses or offices. The transport system is heavily influenced by public policies. For example, governments decide where and when to build new infrastructure, even in countries that have privately owned infrastructure, such as France and Portugal. Governments set regulations for safety (examples include vehicles, crash worthiness; infrastructure, design; speed limits; persons, alcohol; age of being allowed to drive a car), emission of pollutants (CO_2 and noise) and decide on levies on vehicles and fuels. Governments make dedicated public transport policies.

Policy making in general, and therefore also transport policy making, implies making choices, in case of infrastructure-related policies, examples being budget allocations for infrastructure in general and choices between alternatives for a new road or railway line. Because of all the choices to be made, there is a huge need for

---

This chapter is largely based on van Wee (2011), and van Wee and Rietveld (2013a).

B. van Wee (✉)
Faculty of Technology, Policy and Management, Section Transport & Logistic,
Delft University of Technology, Delft, The Netherlands
e-mail: g.p.vanwee@tudelft.nl

P. Rietveld
Faculty of Economics, VU University of Amsterdam, Amsterdam, The Netherlands
e-mail: p.rietveld@vu.nl

C. Basta and S. Moroni (eds.), *Ethics, Design and Planning of the Built Environment*, Urban and Landscape Perspectives 12, DOI 10.1007/978-94-007-5246-7_7,

ex ante evaluations of choice options. An important question therefore is how to evaluate potential options for future transport projects and policies. (Social) Cost-benefit analysis (CBA) nowadays is a very popular ex ante evaluation method in many countries (Hayashi and Morisugi 2000; Bristow and Nellthorp 2000; Grant-Muller et al. 2001).

One of the important effect categories is safety effects. Although since the early 1970s in most western countries the numbers of fatalities have decreased, despite a huge increase in transport volumes, per 10 million inhabitants, hundreds of people get killed each year in crashes. Consequently, safety effects are important, both from a general policy perspective and from the perspective of the ex ante evaluation of candidate policy options. These options can be transport options in general (such as options for new infrastructure) or specifically related to safety. In CBA, the ex ante evaluation of safety effects is generally based on the so-called willingness to pay of consumers: How much money are consumers prepared to pay for a reduction in risks? This chapter aims to discuss this practice from an ethical perspective.

The remaining part of this chapter is organised as follows: Sect. 7.2 introduces the reader to the current state of the art with respect to CBAs for transport. Section 7.3 explains in more detail how CBA deals with safety effects. Section 7.4 is the core of this chapter and discusses the way safety effects are included in transport CBAs. Section 7.5 discusses the implications of the findings for (spatial) policies. Section 7.6 finally summarises the main conclusions of this chapter and briefly discusses the implications for decision making.

7.2 CBA for Transport: An Introduction

Basically a CBA is an overview of all the pros (benefits) and cons (costs) of a project. These costs and benefits are as much as possible quantified and expressed in monetary terms. Benefits are in general based on consumer preferences.[1] Costs and benefits occur in different years within the time horizon of the CBA. To deal with this, they are presented as so-called net present values, implying that taking into account interest and inflation, it is better to have 1 euro or dollar nowadays than in, for example, 2030. The discount rate is used to express this valuation. Final results are often presented in summary indicators. The main indicators that are presented are the difference between costs and benefits, the return on investment and the benefit to cost ratio. Almost every handbook on transport economics pays attention to CBA in transport (see e.g. Blauwens et al. 2008; Button 2010).

There are three major explanations for the popularity of CBA in the ex ante evaluation of infrastructure projects and its role in decision making – these reasons are the main strengths of the method. The first explanation is that most costs and benefits are well known, the second is that models to forecast demand are generally available and the third is that CBA is a relatively "neutral" evaluation method. We will discuss both explanations in more detail below.

In case of possible future transport policy options, certainly in case of infrastructure options, costs and benefits are quite well known. Investment, maintenance and operation costs can be derived from data from projects constructed in the past, or from tenders. The most important benefits are travel-time savings, both for travellers and freight transport. Models are generally used to estimate the demand of passengers or volumes of goods transport that will benefit from a new project, as well as time savings. Next, the so-called value of time (VOT) is used to express shorter travel times in monetary terms. VOT is higher for business travel and goods transport than for commuting, and leisure travel has the lowest value of time. VOT differs between modes, income classes and some other characteristics of travel and travellers (e.g. Gunn 2001). Note that the travel-time savings, often being the most important benefits of infrastructure projects, are not fully expressed in GDP. Travel-time savings for business trips and goods transport lead to higher productivity and lower costs and have an impact on GDP, but if a commuter can leave home later because commuting times are reduced, or because it takes less time to travel to a relative, GDP is not affected. In CBA, it is common to have a broad approach for welfare, implying that all benefits for consumers are included, even if they are not incorporated in GDP.

The second reason for the popularity of CBA is the general availability of models for demand forecasts. The overview of CBA relevant impacts is generally based on state-of-the-art methods, modelling being the method of preference in most cases. In general two categories of models are used, the first one being transport models and the second being impact models (emissions, safety). Many western regions and countries have state-of-the-art models available, at least transport models, but often also impact models, though impact models, especially safety models, are often quite simple – they multiply travel volumes (in general or per mode, or car use only) with risk factors.

The third reason for the popularity of CBA is its often-assumed "neutral" characteristic as opposed to its main competitor: multi-criteria analysis (MCA). In MCA, effects are presented and weighed using weights per effect. Setting the weights is not at all value free. It is therefore much easier to manipulate the final outcomes of an MCA compared to a CBA.

Despite these strengths, several weaknesses exist. It is beyond the scope to give a full overview. Briefly summarising important weaknesses relate to the quality of cost estimates, in some case, the quality of travel demand forecasts; the difficulty to estimate wider economic effects; the difficulty to monetise some categories of effects, such as effects on nature or the quality of the urban environment; the general ignorance of distribution effects (who gains, who loses?); and the poor way of communication of results to nonexperts.

CBA is aimed to allow for a welfare evaluation. The focus on welfare implies a utilitarian perspective. Utilitarianism is the most popular theory of a family of ethical theories called consequentialism. Utilitarianism is a theory within the wider family of consequentialism. Consequentialism "is the view that normative properties depend only on consequences" (*Stanford Encyclopedia of Philosophy*). Utilitarianism, more specifically act consequentialism, "is the claim that an act is morally right if, and only if, that act maximizes the good, that is if, and only if, the total amount of good for all, minus the total amount of bad for all, is greater than this net amount for any incompatible act available to the agent on that occasion" (*Stanford Encyclopedia of Philosophy*). Utilitarianism provides an ethical foundation of CBA: A CBA compares policy options from the perspective of utility. The utility of distinguished effect categories is mainly based on the willingness to pay (WTP) of consumers.

7.3 Safety in CBAs

Also monetary valuation of changes in safety levels is based on the willingness to pay (WTP) of consumers for lower risk levels. A concept that can often be found in monetary valuations of safety effects is the value of a statistical life (VOSL). In economics and in the transport and safety community, monetary valuation of risk changes is much more common than in some other areas, such as in the health sector. An often used "solution" to (potential) moral criticism to express safety in monetary terms is to not price lives directly, but to use what is called a "statistical life". The result is the VOSL, also abbreviated in literature as VSL. The VOSL is an "anonymous indicator" that is used to place a monetary value upon a change in the estimated number of fatalities in traffic over a certain period of time under given circumstances. In fact, a value is put on risk, and this risk is multiplied by traffic or travel volumes. Proponents argue that people may not be able to say how much their life is worth, but they are able to say how much they are prepared to pay for lower risks. And if we know the WTP for lower risks, we have an indication of the VOSL. Examples of choices that people make that give an indication of the WTP could be the safety features in cars (such as additional airbags) or preferences with respect to driving speed (and related travel times).

This implies that the term VOSL is actually somewhat misleading. Its essence is that it represents the valuation of people exposed to travel risks in terms of an amount of euros and dollars (or other currency) per unit of risk reduction. Hence, the use of the VOSL concept is nothing more than a handy way to represent consumer's preferences for risk reductions. The standardisation to the willingness to pay per 1 statistical life saved is, strictly speaking, superfluous. An obvious advantage of the concept of the VOSL is that it is easy to use for communication purposes. The essence of the economical approach is not "pricing human lives", but "pricing human risks".

7.4 Discussion on Current Practice of Including Safety in CBA

Despite its popularity, CBA is often criticised. van Wee (2012) gives an overview of the criticisms as found in literature. Here, we limit the discussion to the inclusion of safety effects in CBA. van Wee and Rietveld (2013b) discuss the current practice of including safety effects in CBA. More specifically, they discuss the following questions:

1. Is it morally acceptable to express (prevention of acceptance of) fatalities or risks in monetary terms?
2. How useful is the concept of the value of a statistical life (VOSL) for ex ante evaluations of transport policy options?
3. What are the pros and cons of expressing (prevention or acceptance of) fatalities or risks in monetary terms in ex ante evaluations?
4. Which methods are available for expressing (protection of) human lives in monetary terms, and what are the main related methodological discussions?
5. Are all safety-related costs generally included in ex ante evaluations of the safety impacts of transport policy options, and if not, what is the relevance of excluded costs categories from an ethical perspective?
6. How important is the distribution of safety effects from an ethical perspective?

They discuss these questions from the perspective of transport safety but state that the discussion might also be relevant for other areas of application, such as risks in industry – see, for example, Evans (2009) who reflects on this topic. Here, we present a summary of the discussions.

7.4.1 Is Pricing Risk Changes Acceptable?

Within the community of persons involved in ex ante evaluations of transport plans and policies, the subject of pricing human lives is one of the most controversial. Some think intuitively that it is immoral to price human beings, others highly support doing this. We will not give a clear and indisputable answer to the question in the heading of this subsection: There is no clear answer. The answer depends on the ethical theory one uses as a basis. The two most extreme positions probably follow from the Kantian perspective, an influential perspective in a category of ethical theories called "deontology" and the utilitarian perspective, an influential perspective in a category of ethical theories called "consequentialism".

Deontology is a category of ethical theories regarding which choices are morally required, forbidden or permitted. In other words, deontology falls within the domain of moral theories that "guide and assess our choices of what we ought to do" (*Stanford Encyclopedia of Philosophy*). Deontologists hold that at least some fundamental moral principles, rules or ideas are to be followed, regardless of the outcomes. The most well-known deontologist is Immanuel Kant. In this chapter, we firstly discuss the Kantian perspective because of its "extreme" position. Kant developed the so-called principle of the categorical imperative. We cite Audi's (2007) formulation because it is much simpler than the original formulation by Kant: "Act in such a way that you always treat humanity, whether in your own person or in the person of any other, never simply as a means, but always at the same time as an end". This applies to oneself as well as to others. Everyone matters and matters equally. Although a Kantian perspective does not directly relate to risks, one could argue such a perspective could lead to the conclusion that pricing the value of a human life is immoral.

People should not be considered as a means to something else. It would then follow, for example, that shorter travel times cannot compensate for a reduction in safety: Travel-time savings should not come at the cost of additional risks for others. This perspective can, but does not necessarily have to, be derived from a Kantian perspective. When considering transport projects and policies, a potential problem with the Kantian perspective is that the ex ante evaluation of transport projects and policies relates to *statistical* lives; there are no clearly defined people who will lose their lives. We consider probabilities, not specific individuals.

7.4.2 The Utilitarian Perspective

We have introduced utilitarianism above. A utilitarian perspective on valuing safety holds that people may not necessarily price their own life or the life of others, but they do value risks. For example, a person buying a new car considers the safety level (at least the perceived level of passive safety, the crash worthiness of a car, as expressed in the Euro NCAP ratings) but also many other characteristics of a car, such as price, size, performance and emotional values. And people know that driving 120 km/h is less safe than driving 100 km/h, but they trade-off travel time, and maybe the fun of driving, and safety (and fuel costs). So maximising any form of utility should include safety. And this can be done, based on the preferences of humans as consumers or – more generally – persons making choices.

Some have even claimed that striving for maximum safety levels is unethical. An example is provided by the discussion on the Swedish Vision Zero (for road traffic). One of the architects behind the vision, Claes Tingvall, stated that the requirements of the vision were so strong that "whenever someone is killed or seriously injured, necessary steps must be taken to avoid a similar event" (Tingvall and Haworth 1999).[2] This position is claimed to be naïve, overly ambitious and even unethical (Fahlquist 2006). Elvik (1999) asserts that the aim to eliminate road traffic deaths would demand such substantial resources that other areas where people's lives are also at risk would suffer.[3]

7.4.3 Deontology: The Doctrine of Double Effect

In addition to the two most extreme positions with respect to pricing changes in safety levels, we now discuss one of several "in-between positions". The position is based on the deontological principle of the "doctrine of double effect", according to which there is a moral difference between causing harm or evil as an unintended side effect of an intended action or policy, and intending the harm of evil directly, either as an end or as a means to an end. From this perspective, it would at least make a difference if a fatality resulted from immorally risky driving behaviour by someone who deliberately endangers the life of others or from "normal" driving behaviour. People who drive riskily do not intend to harm others (or themselves), but they accept endangering others. Examples include drunk driving and speeding. In general, a follower of deontological principles would be reluctant to price accident risks, although the introduction of the distinction between intended and unintended effects according to the "doctrine of double effect" might lead to a refinement of this position. This refinement could be relevant if the position is linked to policy because the two kinds of effects may well differ in their preventability by policy interventions. One could argue, however, that from the perspective of consequentialism, no difference should be made between a fatality resulting from immoral driving behaviour and a fatality resulting from a "normal" accident.

To summarise, there is no clear answer to the question of whether pricing human lives is right. One could wonder: Why discuss the question at all? Our answer would be that we should respect the positions of people who support both sides of the argument.

7.4.4 How Useful Is the Concept of the Value of a Statistical Life (VOSL) for Ex Ante Evaluations of Transport Policy Options?

Above we have introduced the VOSL. This is a seemingly simple concept: Ask for peoples' WTP and derive monetary values for risk reductions from their answers. However, applying the concept is not completely straightforward, a first reason being that the concept of the VOSL (and certainly how it is used in practice) assumes no relationship between the VOSL and risk levels. However, the level of risk can have an impact on the VOSL because it has an impact on choices and WTP. Generally speaking, the higher the risk, the higher the monetary value people put to a constant unit of output (Morton 1991; Hammitt 2007). For example, people might be willing to accept a risk of 1/10,000 for 200 dollars (resulting in a VOSL of two million dollars), but only very few people will accept a risk of 50 % for 4 million dollars. Therefore, the risk level can cause variations in the outcomes of choices, leading to other VOSLs. In other words, the patterns of choice, and thus also the VOSLs, are *risk level dependent*. Due to the increase in VOSL with increasing risk levels (the VOSL could approach infinity for risks nearing the value of 1), it can even be argued that the VOSL is not the same as the value of an unidentifiable person's life (Hammitt 2007).

Secondly, if the VOSL is based on people's choices, it assumes a "correct" perception of the risks. The usefulness of VOSL could be questioned if it is based on a misperception of the risks by people. For example, from research, it is known that people are not very able to deal with very small risks and to translate a small risk reduction into a monetary benefit (Kahneman and Tversky 2000; Kahneman et al. 1982, 1999; Kahneman and Knetsch 1992; De Blaeij 2003). VOSL based on WTP for lower risks may be primarily useful if risk levels in a particular ex ante evaluation matches those of the cases of the stated or revealed preference research. Below we further discuss the problem of discrepancies between objective and subjective safety levels.

Thirdly, the VOSL generally includes a valuation of the statistical lives of the people involved, but not their descendants, though there could be a good point in doing so (see Broome 2005; the explanation is quite complicated and is beyond the scope of this chapter). This point refers to the conventional way of doing WTP research in this area – questions do not explicitly relate to descendants that could be born in the future but are not due to fatalities. A well-designed research could nevertheless include related questions.

Fourth, if one argues that each person is equally important and thus the VOSL (at least within one country) should not depend on income levels, an inconsistency can occur between the VOSL and the value of time (VOT), or more specifically the marginal value of travel-time savings (MVTTS). Below we will use the term VOT. This is because the VOT is income dependent: VOT for higher income groups is higher than for low income groups. The inconsistency only occurs in case an income-dependent VOT would be used. As a result, high income groups score better in a CBA than lower income groups (e.g. Mackie et al. 2003). As a reaction, the so-called "equity value of time" was introduced in virtually all CBAs carried out in the USA and abroad. The equity value of time is based on an average income level (Morisugi and Hayashi 2000).[4] Note that our discussion relates to evaluations, not to forecasting behaviour. In the latter case, person-dependent VOTs and risk acceptance levels should be used.

7.4.5 What Are the Pros and Cons of Pricing (Prevention or Acceptance of) Lives or Risks in Case of Ex Ante Evaluations?

The consideration that pricing of risks is unethical is an extremely powerful argument, probably strong enough to overrule all other arguments for those who accept it. On the other hand, there may be good reasons to value human lives. One reason is that people trade-off safety levels and other impacts of their choices anyway, so using a monetary value for the changes in risks (and, multiplied with volume indicators, resulting in the number of fatalities) based on peoples preferences contributes to a balanced way of including peoples preferences in ex ante evaluations.

A second, and related, reason could be that if a CBA is carried out anyway, outcomes of interest that are not expressed in monetary terms probably have less impact on decision making. Many researchers and policy makers have the impression that decision makers, certainly in the case of a CBA, primarily look at financial indicators such as benefits minus costs, the benefit to cost ratio or the return on investment. If safety is not included, its impact on decision making could be less compared to including safety in monetary terms. Hills and Jones-Lee (1983, p. 355) reflect on the risks of inconsistency and allocative efficiency: "If inconsistency and allocative inefficiency are to be avoided, then explicit monetary costs of accidents and values of accident prevention are required". An equal amount of discussion can also be found in the literature on intergenerational justice: Risk reductions come at a cost. The price of risk reductions to zero can easily be too high (Davidson 2009).

7.4.6 *Which Methods Are Available for Expressing (Protection of) Human Lives in Monetary Terms, and What Are the Main Related Methodological Discussions? An Overview of Methods*

The current state of the art in valuing changes in safety levels is that both material and nonmaterial costs should be included (De Blaeij et al. 2003). Material costs include damage to vehicles and in some cases also infrastructure, loss of production of people and costs of medical treatment. Immaterial costs include loss of the quality of lives of the victims and the people who care about them (family, friends, others).

Several methods are available to express protection of human lives in monetary terms. A first distinction can be made between consumer-based and other methods. In the CBA community, the impression is that consumer-based preferences are generally to be preferred: Who else is better able to value the importance for consumers than the consumers themselves? We will discuss related methods below. Nevertheless, it is good to realise that other methods also exist. Such methods are labelled as "costs per life saved methods" (De Blaeij 2003). Such methods mainly look at choices of policy makers in the past: One can look at the implicit VOSL that results from policy measures taken in the past. This could be done within or outside the transport system. Examples in the transport system could be safety regulations to reduce the risks of car or rail accidents; examples outside the transport system could be regulations for safety standards in industry.

In the category of consumer-based methods, a distinction can be made between willingness to pay (WTP) and willingness to accept (WTA) methods. WTP relates to the willingness to pay for any improvement, such as a reduction in risk levels. WTA relates to the willingness to accept any losses, such as an increase in risk levels, for example, due to other car drivers driving faster. For several reasons, WTP values are of more use than WTA values. One of the reasons is that WTA values that follow from questionnaires could be biased because of strategic behaviour of respondents. In addition, WTA can more easily be nonrealistic: People might (implicitly or explicitly) suggest that they have a much higher WTA measures that will yield risk reductions than they are really prepared to pay for given the choice (see e.g. Hanemann 1991; Perman et al. 2003).

7.4.6.1 A Discussion of Methods

Below we will discuss these methods using ethical theory. An important issue in the valuation of safety is the question: Does each fatality have an equal value or not? If one answers this question with "yes", the conclusion would be that it is as bad if a 90-year-old blind and deaf person dies 2 weeks earlier than he or she would otherwise have done due to high concentrations of ozone, as when a 15-year-old school child gets killed in a road accident. Only a few people would agree. If the answer is negative, the question is: which method to use? There are at least two options: the

WTP as discussed above and the concept of QALYs (quality adjusted life years). We first discuss this concept, followed by a discussion on the inconsistencies between QALY and the WTP-based VOSL.

The concept of QALY is introduced to express the combination of quality and quantity of lost life years. The concept is widely applied in health-care decision making (Loomes and McKenzie 1989). The QALY concept firstly has the advantage that the quantity and quality of lost life years do count. A second advantage is that it can also be used to include injuries causing permanent negative health impacts: Even if the quantity of life years of an injured person remains the same, the loss of quality can be expressed. A major point of discussion is that it is disputable whether very young persons should be compared with others. To quote Morton (1991, p. 112), "for very few people would think that, for example, one should sacrifice more for the safety of a newborn baby than for that of a fifteen-year-old child".

We now continue the discussion comparing QALYs and the WTP. It is important to realise that an inconsistency can occur between WTP and the concept of QALYs. A good overview of the discussion on WTP versus QALYs can be found in Hammitt (2002). He states that although both methods are based on individual preferences, the underlying assumptions differ. The different bases yield systematically different conclusions about the relative value of reducing health and mortality risks to individuals that differ in age, health conditions, income and other factors. The choice of which method to use depends on judgments about what constraints should be placed on individual preferences and what factors should be considered in aggregating preferences across people. Estimates of QALYs are likely to be less variable across people and studies than estimates of WTP because the QALY framework imposes greater constraints. To quote Hammitt (2002, p. 998), "QALYs impose substantial and somewhat unrealistic constraints on the form of individual preferences and combine preferences across people on a relatively egalitarian basis. In contrast, WTP imposes few constraints on individual preferences and gives relatively greater weight to more affluent sectors of society".

The inconsistency between QALYs and WTP firstly relates to the relationships between age and both concepts. Researchers have explored the relationship between the WTP for a statistical life and age. For example, Shepard and Zeckhauser (1984) found that VOSL peaks near age 40 and is less than half as large at ages 20 and 65. The increasing VOSL between the age of 20 and 40 contradicts the QALY concept. An explanation may be that between the age of 20 and 40 income increases. But probably even a person having an increased WTP between the age of 20 and 40 might prefer to get killed in a road accident at the age of 40 rather than at the age of 20. So the inconsistency is notable. De Blaeij (2003) found another example of such an inconsistency between the QALY approach and the VOSL approach, stating that the VOSL peaks between 50 and 65, and hence only starts to decline beyond 65 years. In De Blaeij's analysis, a correction was applied for income levels; hence, the age pattern was not distorted by age-income interrelationships.

The inconsistency between QALYs and WTP next relates to travel mode. For the concept of QALYs, it does not matter how people are killed in a road accident, but for WTP, it may matter. This is illustrated by Johansson-Stenman and Martinsson

(2008) who did research on people's general *ethical preferences* and the value of life. They combine age and mode. The results reveal not only a strongly decreasing "ethical preference value" of a life with age (giving support for the concept of QALYs over WTP); in addition, it shows that pedestrian fatalities are valued higher than fatalities of an equivalent driver.

A third potential inconsistency between QALYs and WPT arises with respect to children. Children hardly have any money, so their WTP for reduced risks will be very low. One can seriously debate if even doing research into the WTP of children for risks is morally acceptable. One could argue that what then matters is the WTP of their parents (Leung and Guria 2006). But suppose a 10-year-old child lost her parents. Would that mean that the WTP for risk reduction of that child is hardly more than zero? Would it make the life of the orphan child of less value than the life of her friend that still has both parents? And if the parents still live, the WTP for reduced risks of children may very well depend on income as well as on the number of children they have. Would this really matter? Many people would feel uncomfortable answering these questions with "yes".

The fourth inconsistency between QALYs and WTP results from the distinction between risky behaviour and people in general. Again this distinction is not relevant for the QALY concept, but it may be for WPT. In literature, a distinction is made between people who utilise substances that are bad for their health, such as smokers and users of illicit drugs, versus people in general. Researchers have found several differences (Johansson-Stenman and Martinsson 2008). People think that persons responsible for their own bad health should be given lower priority (e.g. Anand and Wailoo 2000, Cookson and Dolan 2000). Accordingly it could be that the public's opinion on WTP for risk reductions is lower in case of people with risky behaviour, compared to the wider public.

The fifth inconsistency between both concepts follows from a distinction between involuntary risks versus voluntary risk. This distinction is not relevant for QALYs, but it is for people's ethical preferences and related WTP. People think involuntary risks are to be valued higher (e.g. Slovic et al. 1985; Mandeloff and Kaplan 1989).

Sixth, people think that risks that are difficult to avoid should be valued higher than those that are not (e.g. Subramanian and Cropper 2000), a distinction that is not relevant from a QALY perspective.

Note that several of the differences in ethical preferences do not match the utilitarian perspective taken in a CBA; from a utilitarian perspective, it would not matter whether, for example, a person gets killed in an accident that was easily avoidable or not. But CBA would favour spending on effective measures to prevent avoidable accidents overspending on vain attempts to prevent unavoidable ones.

It has to be added that the above discussion on the valuation of life risks of different categories of people is somewhat theoretical if it is compared with real-world applications of the VOSL in the cost-benefit analysis of transport policies. Real-world applications tend to avoid the use of differentiation values for different types of people at risk and just apply an average value. There are probably two reasons for this. First, the overall quality of estimates of VOSL is probably not strong enough to allow specific values for various subgroups. Second, the researchers responsible for the cost-benefit calculations may fear debates about "unethical" assumptions on which the calculations are based. They prefer therefore to stay on the safe side by using just the average VOSL. A similar reason would be that researchers doing CBA would anticipate that the application of strongly differentiated VOSL levels might lead to conclusions that decision makers might find difficult to swallow, like a low priority for traffic safety themes that would in particular benefit children. Thus, by using an average value for the VOSL, analysts responsible for CBA make sure that the potential gap between market-oriented economics-based policy support on the one hand and the domain of policy convictions and equity concerns on the other hand can be kept to a manageable size.

A second subject for methodological discussion is discounting future safety effects. Discounting reduces the value attributed to long-term benefits. Applying usual discount rates to protection of human lives would reflect a preference for preventing a person's death now over preventing the death of an equivalent person in the future. This makes perfect sense in the context of CBA. Consider, for example, the case that an investment has safety benefits now compared with an equally expensive investment that has identical safety benefits in the future. Then it makes sense to prefer the investment with immediate effect over the one with delayed effect. Along similar lines, not discounting the benefits of investments in terms of lower probabilities of death (or higher QALYs), while at the same time discounting the costs of such measures, leads to the implausible result that postponing this investment is always to be preferred (e.g. Keeler and Cretin 1983; Hammitt 2002). The reason is that postponement would lead to lower costs given the discounting, whereas the benefits would remain unaffected. Thus, the case for discounting VOSL or QALYs is stronger than one might think. Johannesson et al. (1994) indicate that a limited change in definition – and measurement – of QALYs would suffice to allow discounting within this concept.

A third subject of methodological discussion is related to the fact that objective safety and perceived safety do not always match. Traffic situations can be unsafe, but people do not always perceive that to be so. Alternatively, objective numbers show that risk levels are low, but people might feel unsafe. Research has shown that the correlation between objective and subjective safety is often poor, not only in the area of road safety (Vlakveld et al. 2008) but also elsewhere in society (Nilsen et al. 2004). Does this matter? From a CBA (utilitarian) perspective, the answer could be that it matters if people are willing to pay (WTP) for an increase in perceived safety *even if the objective safety does not change by an equal amount*. This subject can be seriously debated. People might be prepared to pay for increased safety but only if objective safety increases. If they initially thought that a traffic situation would become safer, but they find out – or are informed – later that it does not, one can seriously doubt the WTP for increase perceived safety. In fact, one can doubt if perceived safety will increase at all if people know the real safety levels.

A fourth subject for methodological discussion is the interaction between risk influencing factors. In the case of road safety, risk levels result from (at least) driving speed, the use or otherwise of protective devices of cars (active and passive safety levels), infrastructure characteristics and the quality of the health-care system.

In addition, some of these determinants interact, and this interaction has an impact on the final risk levels. For example, if cars become safer, drivers may drive faster. But the combined impact on risk levels may be infrastructure dependent. Understanding such interactions is primarily a challenge for researchers. But an important ethical question is whether changes in determinants and their interactions should firstly have design consequences and secondly consequences for evaluating designed options. Should, for example, safer cars lead to higher maximum speeds at motorways? Calculations based on the "optimal" design of speed on motorways from a utilitarian perspective would argue so. On the other hand, higher speeds can lead to changes in the distribution of risks. For example, low income people may have smaller, less safe cars and drive at lower speeds compared to high income people with big, new, safer cars. Would that matter? Egalitarian theories would argue that it does; consequentialism would probably conclude that it does not.

Finally, a methodological discussion is the transferability of results over time and space. To start with the transferability over space, it can be highly ethically problematic to transfer the outcomes of one country to another country. For example, the VOSL based on WTP in the USA will not be of value for evaluating the lives of people in Bangladesh, and vice versa. For a discussion of the impact of the world region under consideration on the ethics of fatalities, see Lorenzo et al. (2010). In addition, the transferability of VOSLs over time deserves attention. If people get richer, their WTP for risk reductions is likely to increase. But are the lives of rich people of more value than those of poor people? In other words, is the WTP the best method for valuing a statistical life? On the other hand, an ageing population would – ceteris paribus – lead to a decrease in the VOSL if it was based on the concept of QALYs.

7.4.6.2 Are All Safety-Related Costs Generally Included in Ex Ante Evaluations of the Safety Impacts of Transport Policy Options, and if Not, What Is the Relevance of Excluded Costs Categories from an Ethical Perspective?

We argue that a certain category of avoidance costs is missing in CBA. We firstly introduce the concept of avoidance costs, followed by a discussion of a specific category of avoidance costs that is generally missing in the societal ex ante evaluation of policy options.

Avoidance costs are costs made to improve safety and can be split into several categories:

1. Infrastructure-related costs
2. Vehicle-related costs
3. Costs related to the health system
4. Costs related to changes in human behaviour

The first three categories are usually included in CBA and will not be discussed here. The fourth category of avoidance costs is the costs of changes in human behaviour due to (perceived) changes in safety. People can adapt their behaviour because they perceive safety levels to be low. For example, older persons may prefer to stay at home because they think travelling is too risky. Or they may travel by taxi because they perceive cycling to be too risky. Or a person may prefer to cycle but travels by car because of a perceived low safety level of cycling. In addition, parents may not allow their children to travel to school independently because of a certain (perceived) risk and therefore bring their children to school themselves. Or they may want them to use the school bus instead of cycling to school. Such adaptations come at a cost. In addition, if adaptations result in a decrease in the use of slow modes, there are losses related to health. For example, the health benefits of cycling are substantial. In a Norwegian study on costs and benefits of cycling infrastructure in cities, these benefits count for more than half (55–75 %) of all benefits of cycling (Saelensminde 2004).

So at least theoretically, avoidance costs are relevant for ex ante evaluations. But are they in practice? We think in many cases the *changes* in perceived risks due to candidate policy options are very low, and consequently, the impact of ignoring avoidance costs may be small. But in some cases, they could matter, examples being changes in maximum speeds on distinguished road classes or the planning of schools and related routes between homes of school children and schools. In addition, they are relevant when an estimation of the total costs of safety needs to be made.

7.4.6.3 How Important Is the Distribution of Safety Effects from an Ethical Perspective?

It is very possible that the pros and cons of policy options related to safety in the transport system are not equally distributed across the population. This distribution is relevant from an ethical perspective. Trade-offs may exist between car users and others (e.g. children, elderly who do not drive). Such trade-offs exist in both directions. It is the non-car user who benefits from restrictions with respect to car use at the cost of car users. If priority is given to car users, the latter benefit, at the cost of the non-car users. If and how distribution-related impacts should be evaluated depends on the ethical perspective. CBA and a related utilitarian perspective would allow for a straightforward calculation of utilities, either simply summarised in the value of a single indicator or accompanied by an estimated distribution of benefits and disbenefits over various categories of affected (groups of) people. However, egalitarian theories would specifically address distribution effects. A focus on, for example, the 20 % of people who are "worse off" in the transport system would probably result in a shift to policies that favour the safety of the non-car user. Distribution effects also matter from the perspective of sufficientarianism, which holds the view that what primarily matters is that everybody is well-enough off, that is, has well-being above a certain given threshold which is considered "sufficient". For "weak sufficientarianism", the improvement of well-being matters if people's well-being is below a threshold. The lower the level of well-being, the higher the moral value of benefiting a person. For " strong sufficientarianism", absolute priority should be given to the improvement of well-being of those whose level of well-being is below the threshold. And the lower their welfare, the more important it is to improve their well-being (Meyer and Roser 2009; see also Wolf 2009). The perspective of strong sufficientarianism could even imply that absolute priority should be given to improving safety if safety levels are below the minimum level. A problem then exists that a sufficientarianism approach relates to persons, not to (segments of) infrastructure or vehicles, whereas safety policies often do not focus on individuals, although traffic education, driving lessons and obligations like wearing helmets and not drinking are exceptions. Safety policies often try to make infrastructure or transport modes safer. In such cases, benefits are distributed in a rather diffuse manner which makes it difficult to link them to specific individuals, so that overall safety levels at the individual level cannot be estimated. The sufficientarianism approach implies "personalising" safety, and therefore, bringing this approach into practice is not at all straightforward and needs to create stronger links between people and safety relevant policy options. To conclude, in ex ante evaluations of the safety impacts of policy options, the indicator chosen can easily lead to overlooking such ethically relevant impacts on distribution, but need not do so.

7.5 Implications for Spatial Policies

Spatial policies relate to land-use policies and (often related) infrastructure policies. Such policies in general can have safety impacts. Firstly, land-use policies may have safety implications. For example, planning a school at a cheap location, but at the "wrong side" of a risky road (from the perspective of the residential area where most children live), results in cost savings, but in higher risks. And densification and mixing land-use categories reduce passenger transport in general (less passenger kilometres) and increase the share of slow modes, and such travel behaviour changes may have safety implications. Infrastructure policies relate to which infrastructure is decided upon, at which locations, and with which design characteristics, and which regulations (e.g. speed, overtaking) apply. Such policies may have important safety implications.

In addition to spatial policies in general, these can also be applied to design the built environment in such a way that safety effects are reduced. Infrastructure examples include safe pedestrian crossing, separate lanes for cyclists and speed bumps to reduce speed of motorised traffic. Examples of land-use policies include planning mixed use at the neighbourhood level to avoid pedestrians and cyclists having to use risky roads and the planning of offices near stations to increase the share of the relatively safe train, at the cost of the more risky car. Land-use policies may also be tailored to reduce non-transport-related risks such as third-party risks of manufacturing industry, for example, by creating buffer zones between those industries and residential areas – such policies are quite common in many countries.

So, the evaluation of such policies, and next the use of the evaluations for policy making and decision making, may need careful consideration of how to include safety effects. Based on the discussions of this chapter, we argue that the use of the WTP for risk reductions could provide a first way to evaluate effects. But we think a careful consideration from both a methodological and an ethical perspective according to the lines of this chapter is recommended.

7.6 Conclusions: The Link Between Ex Ante Evaluations and Policy

In our opinion, it is important, if ex ante evaluations are used as input for decision making, that the research is of the highest quality. A practical rule of thumb could be that the quality of this research is higher if as a result of it, the decision makers make the choice they would have made (i) if they had all the potential choice options available, (ii) if they were fully informed and (iii) if they were able to evaluate different choice options. Giving more information on the value of risk improves the information base of decisions and hence would increase the quality of decision making according to the above rule of thumb. We can therefore carefully conclude that expressing safety in monetary terms more often increases the usefulness of research for decision making than that it decreases. A possible exception may be that information overload may undermine the quality of decision making (see Knockaert et al. 2010). However, when cost-benefit information is presented in a proper way with different options in terms of details on certain cost and benefit items, this is probably not a real problem. Another exception might occur when the additional information provided – although being correct as such – is misleadingly presented and thus not well understood by the user. This is indeed a point of interest, implying that the information provided must minimise the risk of misinterpretation.

Notes

1. In some cases, benefits are not based on consumer preferences. Examples include the valuation of CO_2 emission; current consumer preferences are generally much lower than estimates based on policy and political choices.
2. Cited in Elvebakk and Steiro (2009).
3. Cited in Fahlquist (2006).
4. Cited in Martens (2006).

References

Anand P, Wailoo A (2000) Utilities versus rights to publicly provided goods: arguments and evidence from health care rationing. Economica 6:543–577

Audi R (2007) Moral value and human diversity. Oxford University Press, Oxford/New York

Blauwens G, De Baere P, Van de Voorde E (2008) Transport economics. De Boeck, Antwerp
Bristow A, Nellthorp J (2000) Transport project appraisal in the European Union. Transp Policy 7(1):51–60
Broome J (2005) Should we value population? J Political Philos 13(4):399–413
Button KJ (2010) Transport economics, 3rd edn. Edward Elgar, Cheltenham
Cookson R, Dolan P (2000) Principles of justice in health care rationing. J Med Ethics 26:323–329
Davidson MD (2009) Acceptable risk to future generations. In: Asveld L, Roeser S (eds) The ethics of technological risk. Earthscan, London/Sterling, pp 77–91
De Blaeij AT (2003) The value of a statistical life in road safety; Stated preference methodologies and empirical estimates for the Netherlands. Tinbergen Institute Research Series & Vrije Universiteit, Amsterdam
De Blaeij AT, Florax RJGM, Rietveld P, Verhoef E (2003) The value of statistical life in road safety: a meta-analysis. Accid Anal Prev 35(6):973–986
Elvebakk B, Steiro T (2009) First principles, second hand: perceptions and interpretations of vision zero in Norway. Saf Sci 47:958–966
Elvik R (1999) Can injury prevention efforts go too far? Reflections on some possible implications of Vision Zero for road accident fatalities. Accid Anal Prev 31(3):265–286
Evans AW (2009) Economic appraisal of road safety measures in Great Britain. Int J Sustain Transp 3(3):160–177
Fahlquist JN (2006) Responsibility ascriptions and vision zero. Accid Anal Prev 38:1113–1118
Grant-Muller SM, MacKie P, Nellthorp J, Pearman A (2001) Economic appraisal of European transport projects: the state-of-the-art revisited. Transp Rev 2(2):237–261
Gunn H (2001) Spatial and temporal transferability of relationships between travel demand, trip cost and travel time. Transp Res E 37:163–189
Hammitt JK (2002) QALYs versus WTP. Risk Anal 22:985–1001
Hammitt JK (2007) Valuing changes in mortality risk: lives saved versus life years saved. Rev Environ Econ Policy 1(2):228–240
Hanemann M (1991) Willingness to pay and willingness to accept: how much can they differ? Am Econ Rev 81(3):635–647
Hayashi Y, Morisugi H (2000) International comparison of background concept and methodology of transportation project appraisal. Transp Policy 7:73–88
Hills PJ, Jones-Lee MW (1983) The role of safety in highway investment appraisal for developing countries. Accid Anal Prev 15(5):355–369
Johannesson M, Pliskin JS, Weinstein MC (1994) A note on QALYs, time tradeoff, and discounting. Med Decis Mak 14:188–193
Johansson-Stenman O, Martinsson P (2008) Are some lives more valuable? An ethical preferences approach. J Health Econ 27(3):739–752
Kahneman D, Knetsch JL (1992) Valuing public goods: the purchase of moral satisfaction. J Environ Econ Manag 22:57–70
Kahneman D, Tversky A (2000) Choices, values and frames. Cambridge University Press, Cambridge
Kahneman D, Slovic P, Tversky A (1982) Judgement under uncertainty: Heuristics and biases. Cambridge University Press, Cambridge
Kahneman D, Ritov I, Schkade D (1999) Economic preferences or attitude expressions? An analysis of dollar responses to public issues. J Risk Uncertain 19:203–235
Keeler EB, Cretin S (1983) Discounting of life-saving and other nonmonetary effects. Manag Sci 29:300–306
Knockaert JSA, Evangelinos C, Rietveld P, Wieland B (2010) Differentiated infrastructure charging: a comparison of theory and practise. Eur Transp 43:4–34
Leung J, Guria J (2006) Value of statistical life: adults versus children. Accid Anal Prev 38:1208–1217
Loomes G, McKenzie L (1989) The use of QALYs in health care decision making. Soc Sci Med 28(4):299–308

Lorenzo O, Esqueda P, Larson J (2010) Safety and ethics in the global workplace: asymmetries in culture and infrastructure. J Bus Ethics 92:87–106

Mackie PJ, Wardman M, Fowkes AS, Whelan GA, Nellthorp J, Bates J (2003) Value of travel time savings in the UK. Prepared for the Department for Transport. Institute for Transport Studies, University of Leeds/John Bastes Services, Leeds

Mandeloff M, Kaplan R (1989) Are large differences in lifesaving costs justified? A psychometric study of the relative value placed on preventing deaths. Risk Anal 9:349–363

Martens K (2006) Basing transport planning on principles of social justice. Berkeley Plann J 9 (www-dcrp.ced.berkeley.edu/bpj)

Meyer LH, Roser D (2009) Enough for the future. In: Gosseries A, Meyer LH (eds) Intergenerational justice. Oxford University Press, Oxford, pp 219–248

Morisugi H, Hayashi Y (2000) Editorial. Transp Policy 7(1):1–2

Morton A (1991) Disaster and dilemmas. Strategies for real-life decision making. Basil Blackwell LTD., Oxford

Nilsen P, Hudson DS, Kullberg A, Timpka T, Ekman R, Lindqvist K (2004) Making sense of safety. Inj Prev 10:71–73

Perman R, Ma Y, McGilvray J, Common M (2003) Natural resource and environmental economics. Pearson Education Ltd./Ashford Colour Press Ltd., Gosport

Saelensminde K (2004) Cost–benefit analyses of walking and cycling track networks taking into account insecurity, health effects and external costs of motorized traffic. Transp Res A Policy Pract 38(8):593–606

Shepard DS, Zeckhauser RJ (1984) Survival versus consumption. Manag Sci 30:423–439

Slovic P, Fischhoff B, Lichtenstein S (1985) Regulation of risk: a psychological perspective. In: Noll R (ed) Regulatory policy and the social sciences. University of California Press, Berkeley, pp 241–278

Subramanian U, Cropper M (2000) Public choices between life saving programs: the tradeoff between qualitative factors and lives saved. J Risk Uncertain 21:117–149

Tingvall C, Haworth N (1999) Vision zero – an ethical approach to safety and mobility. In: Paper presented to the 6th ITE conference on road safety and traffic enforcement: beyond 2000, Melbourne, 6–7 Sept 1999

van Wee B (2011) Transport and ethics. Ethics and the evaluation of transport policies and projects. Edward Elgar, Cheltenham

van Wee B (2012) How suitable is CBA for the ex-ante evaluation of transport projects and policies? A discussion from the perspective of ethics. Transp Policy 19(1):1–7

van Wee B, Rietveld P (2013a) CBA: ex-ante evaluations of mega-projects. In: Priemus H, van Wee B (eds) International handbook on mega-projects. Edward Elgar, Cheltenham

van Wee B, Rietveld P (2013b) Using value of statistical life for the ex-ante evaluation of transport policy options – a discussion based on ethical theory. Transportation (conditionally accepted)

Vlakveld WP, Goldenbeld Ch, Twisk DAM (2008) Beleving van verkeersonveiligheid; Een probleemverkenning over subjectieve veiligheid. Report R-2008-15. Stichting Wetenschappelijk Onderzoek Verkeersveiligheid SWOV, Leidschendam

Wolf C (2009) Climate change and climate policies. In: Gosseries A, Meyer LH (eds) Intergenerational justice. Oxford University Press, Oxford, pp 347–376

Chapter 8
Urban In/Justice

Emily Talen

8.1 Introduction

What makes a given urban element, in a given location, unjust, as opposed to merely unfortunate? What would injustice be like, in a literal sense?

While there is wide interest in the theoretical aspects of urban justice, I argue in this chapter that not enough time has been devoted to grounding these theoretical explorations in terms of *literal* interpretation, or how urban injustice is manifested in everyday urban experience. I contend that it is not only possible but necessary to make the injustices of everyday urban life explicit and meaningful – beyond the

Suburban wasteland. Photo courtesy of the author

E. Talen (✉)
School of Sustainability and Geographical Sciences and Urban Planning,
Arizona State University, Tempe, AZ, USA
e-mail: etalen@asu.edu

C. Basta and S. Moroni (eds.), *Ethics, Design and Planning of the Built Environment*, Urban and Landscape Perspectives 12, DOI 10.1007/978-94-007-5246-7_8,

abstract and beyond the theoretical. For example, why would the above photo of the built environment be considered "unjust"?

Consider the wide range of built environments that could be interpreted through the lens of "justice": (i) a trailer home in a remote location, (ii) a sidewalk that ends abruptly, (iii) a wide arterial road separating a shopping center and a neighborhood, (iv) vacant land in the heart of the city, (v) cul-de-sac streets, (vi) big box stores, (vii) apartment housing complexes adjacent to industrial land, and (viii) parking garages and parking lots.

Why should any of these elements be interpreted as being unjust, as opposed to being merely objectionable to some people (or even beloved, perhaps, by others)? What is the justification for elevating them to the level of injustice? This chapter lays out a theoretical and practical argument for a literal, physical interpretation of in/justice. I use the term "in/justice" to highlight the fact that injustice and justice in the built environment are often simultaneously exposed. I argue that the failure to translate in/justice – that is, both justice and injustice – more literally is rooted in an overcautiousness concerning past experience (an experience that could be rectified) and, as well, an under-theorizing about what a good city is, or should be.

Often injustice is easier to see and is in many ways in greater need of explanation. Justice, on the other hand, can be translated proactively, as a basis for positive intervention. For example, the built environment can be used to increase socialization, reduce crime, enhance civic pride, promote health, allay the fears that arise from uncomfortable proximities, resolve contestations over space, or balance the problems that ensue when residents have an increased need for privacy and security. In turn, any of those goals could be justified on the basis of connecting them to in/justice.

8.2 Theories of Urban In/Justice

Theoretical work on justice in the city is an essential basis for exploring literal interpretation. Urban geographers like David Harvey and Edward Soja have written prolifically about spatial justice in an urban setting, using critical theories of space and revealing how "unjust geographies" are created. Harvey's seminal *Social Justice and the City* (1973) explored the relationship between social justice and space and called into question the existing "structural limitations" of spatial definition. Harvey called for a move away from technical solutions toward something more subjective and revolutionary.

Susan Fainstein's book *The Just City* (2011) offers an evaluation of the justice of urban policies from an urban planning perspective. The tensions between planning and markets or between efficiency and equity are analyzed. Like many books in this genre, Fainstein faults neoliberal planning policies for promoting economic growth at the expense of social justice. She offers policies and programs to ensure greater justice in both process and outcome. The institutions and social movements that need to be tapped to arrive at a greater level of urban justice are divulged: government programs, redevelopment policies, and an inclusionary, substantive discourse.

There is a large literature that evaluates the meaning of the built environment in sociological, theological, and cultural terms, and much of this work incorporates some interpretations of justice. This includes explorations of what land, public spaces, and buildings mean in terms of theology (Gorringe 2002), sociology (Michelson 1970), architecture (Girouard 1985), and culture (Davis 1999). There are detailed accounts of struggles to control public space, like Mitchell's *The Right to the City* (2003) which advance social justice arguments. A significant literature explores the implications of sprawl and "bad" urban form from economic, social, and environmental points of view. A recent work by Williamson (2010) evaluates sprawl's effect on civic engagement, inequality, and environmental impact.

Most of these writers also consider how a more socially just city might be achieved. Through changes in policy, program, and process, they often target the needs of specific populations. Thus the Lefebvre-inspired group "Right to the City" (a nonprofit organization) is focused on the needs of low-income communities of color, illuminating injustices in the provision of housing, healthcare, and other services. Methods for increasing their participation in decision-making are explored. Environmental justice and spatial justice are translated to practice by involving marginalized groups more directly in participation processes and trying to give them a more powerful voice in urban politics.

There is also a large literature that explores social justice from a strictly spatial/locational point of view. For example, research has focused on the social justice aspects of locating unwanted public facilities (an exploration of "environmental racism"). Researchers have investigated the relationship between minorities and low-income groups and unwanted environmental hazards. For example, Grineski et al. (2007) looked at the class and ethnic environmental injustices of air pollution, attributing the higher exposure of hazards suffered by marginalized populations to white privilege. Promotion of justice in the built environment, then, is a matter of implementing a more equitable distribution of unwanted uses as well as a fairer distribution of desirable resources. Such a distribution might be based on need, matching resources, and facilities to the populations that most need access to them.

Despite theoretical backing, however, the translation of social justice to principles of physical planning and design is weak. I attribute this, in part, to a lack of specificity – an avoidance of translating theory to design principle in anything but the vaguest of terms (there are reasons for this avoidance, which I discuss below). Most often, the action plans of theoretical works on urban justice conclude with platitudes: statements about the need to reduce the urban ecological footprint, elevate local empowerment, meet basic human needs, and promote new forms of governance – all admirable goals. But such declarations tend to be un-actionable. Further, there is often little understanding of what these goals would mean for everyday urban life, or how they would change the specific outlines of urban structural form. Human geographers might study intently "the importance of spatiality in the processes of social reproduction," but the spaces to be studied, the "discourse-producing sites" like prisons, schools, and hospitals, are studied devoid of context. Critical social theory perspectives on design are particularly skeptical of the need to make these kinds of literal connections (see Knox 2010 or Cuthbert 2006).

What is missing, I believe, is a more unequivocal, literal, and physical interpretation of in/justice. For that to occur, it is essential to avoid the claim that the built environment can be used to overcome injustices rooted at a deeper, structural level. The built environment *augments* the programmatic and process-oriented requirements for greater social justice. It does not replace them. Yet the possibility and potential of this kind of complimentary linkage is rarely pursued. Instead, the notion that the physical environment can be crucial in determining social well-being and equality is regularly critiqued. This response, in my view, seems unnecessarily limited in its exclusion of the built environment and its reliance on abstract notions of city form.

What are the underlying reasons for this dismissal? To some extent, the disconnect between theories about urban in/justice and their physical interpretation simply mirrors the loss of localized form as a context for production and consumption – the substitution of "flows and channels" for spatial places (Castells 2003). But there is also the fear that physical form in connection with something as fundamental and overarching as social justice will appear too deterministic and controlling, perhaps implicating what Harvey (1989) calls a "localized aesthetic image" that supports the "capitalist hegemony over space." It may be seen as an attempt to disguise the underlying political and economic processes of injustice, instead motivating "simplistic spatial solutions to complex social, economic, and political problems" (Crump 2002).

Architects, too, are often averse to attaching social agendas to physical urban forms. Attempts to promote social justice through urban design are often limited to innovative approaches to socialized housing. This cautiousness is not unjustified. The application of urban design to social justice has gone badly in the past. The failure of modernist urbanism and its literal-minded articulation of equality in built form resulted in massive demolition of public housing only a few decades after construction. During the urban renewal schemes of the 1950s and 1960s, neighborhood destruction was undertaken in the name of social equality. Given the way in which physical solutions have been cast as cure-alls, critics are right to guard against "bricks and mortar" remedies at the expense of people, institutions, and political process.

Another factor in the lack of connection between in/justice and built form is the weak position of normative theory – that is, theory about what constitutes a good city, something Kevin Lynch lamented in this classic work, *Good City Form* (1981). In the urban planning field, theories about good city form have been subordinated to theories about urban process. Richard Klosterman (2011) recently completed a survey of planning theory courses in US planning schools and found that planning theory mostly consists of critiques of the rational model, perhaps from a gender, space, or postmodern perspective. There is likely to be a strong dose of critical theory and exploration of Habermas' perspective on communicative action as well as Schon's exploration of phenomenology as a way of analyzing planning practice, mediation, and negotiation. Theories related to strategic planning, ethics, and advocacy planning are also likely topics. Planning theory, in short, has remained focused on the procedural side of planning, while the substantive side – theories of what makes a good city– remains undefined. Beauregard's earlier critique of planning theory (1995) showed how limited planning theory had become: "[P]ractitioners have little use for it, students (for the most part) find it a diversion from learning how to do planning and a requirement to be endured, and planning academics, on average, tolerate it. Within academia, planning theory is marginalized; within practice it is virtually ignored."

Lacking a normative theoretical basis, different conceptions of the good city cannot be identified or ranked, and, in general, there can be no ability to decide between different substantive conceptions of what good cities – just cities – are supposed to be. This relativistic approach cannot be countered if planners lack the substantive theoretical content required to argue one perspective over another. In short, theory that is focused on observing and criticizing practice rather than offering a compelling model of good cities stymies the implementation of practical or physical steps toward justice.

In response to the lack of theory rooted in urban form, as well as a planning field insufficiently engaged with physical planning, there have been calls to put the physical realm – what Beauregard succinctly termed "things" – back in planning (Beauregard 2012). More recently, in a related argument, Thomas Campanella (2011) argued for a physically rooted planning field: proactive, visionary, a profession with "disciplinary identity" focused on physical planning, placemaking, and a shared civic realm. Without that reorientation, and lacking theoretical support, the ability to promote justice through physical design will be difficult to advance.

8.3 Literal Interpretations: A Few Examples

If normative theory about good city form could find a more prominent place in urban discourse, how might the literal interpretation of in/justice then be advanced? Theoretical backing is necessary as justification for why a given element, design, or aspect of physical urban form should be valued from an in/justice point of view, but what would the explicit translations be?

Quite simply, it is possible to evaluate the urban in/justice of a given element, form, or design on the basis of *what* it is and *where* it is – its three-dimensional form and its two-dimensional context. The attribution of in/justice is thus an outcome of the intrinsic qualities of the element itself in addition to its spatial location and what surrounds it. For example, inaccessibility to what is desirable can be interpreted as an expression of urban injustice. Proximity to what is undesirable is an expression of injustice.

A wide range of examples of city form have the ability to be interpreted in terms of in/justice. Notions like spatial equity, the meaning and interpretation of access, the importance of diversity, the experience and impact of fear, opportunities for control, sense of place, and the role of community – are all notions of in/justice that have literal outcomes. Other examples that include the cost of vehicle ownership, linkages between the built environment and health, the relationship between street design and traffic accidents, urban heat island effects associated with built forms, housing affordability in relation to built environment, energy use and built form, and crime and street design are all areas in which form and in/justice could be explicitly linked.

These examples are not abstractions. For example, advancing an argument that cul-de-sac street patterns in the suburbs might be interpreted as unjust is supported by empirical work that shows that such patterns are linked to an increase in carbon emissions and lower air quality and increased obesity by reducing walking, or that they contribute to the infeasibility of transit. The injustice of a parking lot might be similarly argued by quantifying associated heat island effects and crime rates.

A range of social effects could be interpreted from an in/justice perspective, in turn advanced or inhibited through urban form. For example, it could be argued that social connectivity is an aspect of urban justice, while social isolation is an aspect of urban injustice. Both are affected by physical design. Increasing connectivity translates to gridded street networks, short blocks, streets that connect rather than dead end, the establishment of central places where multiple activities can coalesce, or the provision of well-located facilities that function as shared spaces. Enhancing connectivity can be as simple as delineating safe places to cross existing streets, calm traffic down on busy avenues, or institute better pedestrian pathways. Less directly, an increase in social connectivity has been found to result from feelings of safety (Newman 1972), from greater utilization of public space (Levine 1986), and from greater use of local facilities for shopping (Riger et al. 1981) – all strongly impacted by city form. "Social seams" in the form of schools, parks, or neighborhood stores have been shown to promote stable, socially diverse neighborhoods (Nyden et al. 1998).

The in/justice implications of health and its connection to the built environment are another important example. A large literature now supports the connection between health, physical activity like walking, and built form (Ewing and Cervero 2010). In the transportation field, neighborhood design has been connected to transport-related physical activity, which has the dual advantage of promoting public health while at the same time addressing transportation problems like congestion, pollution, and greenhouse gas effects (Badland et al. 2008). These issues are matters of in/justice.

Urban in/justice is more readily interpreted as a matter of equalizing access and ensuring closer proximities between where people of all ages and both genders live and work, as exemplified in Dolores Hayden's proposal for a "nonsexist city" (1980). Access to resources defines the "geography of opportunity," where proximity to resources significantly impacts the ability of low-income residents to improve their lives (Briggs 2005). In this sense, the way cities and neighborhoods are designed has a direct bearing on whether access between residents, their places of work, and the services they require is increased or not. Access, in other words, is a form of justice, strongly affected by the built environment. Some forms, such as low-density sprawl, pose a significant barrier when it comes to the provision of neighborhood-level facilities or access to jobs and urban services.

A just city is also a safe city, and physical context plays a strong role. For example, safety may be increased where there is housing integration (i.e., housing that is integrative, not walled off and abruptly insular), options for surveillance, and public places that are active, often through the promotion of intensive use. Safety through design often calls for clear demarcations between public and private space, whereby urban areas have unambiguous functionality. Jane Jacobs (1961) introduced the notion of "eyes on the street" (so it would be relatively effortless for people to keep an eye on neighborhood activities as part of their everyday routines), while Oscar Newman's (1972) defensible space principles included restricting access at certain points. Urban form can increase security by activating "dead" space – empty, unclaimed, or underutilized land for which clear ownership is ambiguous and for which there is little security for the passer by. Alice Coleman's "variable design strategies," Bill Hillier's "space syntax theory," and Christopher Alexander's "pattern language" all call for urban designs that maximize natural surveillance.

These are very practical matters, and perhaps there is little disagreement that justice via access and safety are strongly affected by physical form. Less direct, but potentially just as powerful, are notions about the visual, aesthetic experience of urban dwellers and the impact those experiences have on urban in/justice. Perhaps the case can be made that some visual coherence, some framework for making sense of the urban realm, is an essential basis for a just city. Kevin Lynch's (1981) "dimensions" were aimed at a built environment that could respond to people's needs, including not only access but also vitality and imageability. Could a just city be predicated on vitality, activity, and liveliness, via a physical realm that promotes exchange, social connection, and daily life or presents a positive, culturally meaningful experience?

8.4 Conclusions

In this chapter, through the examples of spatial equity, access, social connectivity, health, safety, and aesthetic experience, I have argued that it is possible to translate ideals about urban in/justice in literal terms. I argued that it was not necessary to constrain notions of in/justice by viewing them abstractly, in platitudes or only in theoretical terms. In fact, the physical implications of urban in/justice are often direct, design based, and actionable. Considering again the photograph that opened this chapter, might the scene depicted be labeled unjust given its lack of pedestrian access, its poor connectivity, its likely effect on physical activity and therefore health, its safety concerns, and its dispiriting aesthetic experience?

Of course, the literal translation of justice and injustice will never be completely straightforward. One person's sense of injustice might be another's sense of justice. Certain factors, once brought in, may trump injustice in favor of other factors, like efficiency or expediency. In the cul-de-sac example cited earlier, it may be possible that increases in sense of community for one group, attributable to the cul-de-sac, are believed to offset the injustices suffered by a wider group. But rather than revert to clichés, we should find a way to consider the complexities inherent in justice definitions, confronting rather than avoiding the manner in which notions of justice, in their physical translation, may conflict. The literal application of urban in/justice may compete with a whole range of political or economic considerations, but, at least, the extent and nature of these connections should be thoroughly considered.

References

Badland HM, Schofield GM, Garrett N (2008) Travel behavior and objectively measured urban design variables: associations for adults traveling to work. Health Place 14:85–95
Beauregard RA (1995) Edge critics. J Plann Edu Res 14(3):163–166
Beauregard RA (2012) Planning with things. J Plann Edu Res 32(2):182–190
Briggs X (2005) The geography of opportunity: race and housing choice in metropolitan America. Brookings Institution Press, Washington, DC
Campanella T (2011) Jane Jacobs and the death and life of American planning. In: Page M, Mennel T (eds) Reconsidering Jane Jacobs. Planners Press, Chicago, pp 141–179
Castells M (2003) The new historical relationship between space and society. In: Cuthbert AR (ed) Designing cities: critical readings in urban design. Blackwell, Oxford, pp 59–68
Crump J (2002) Deconcentration by demolition: public housing, poverty, and urban policy. Environ Plann D Soc Space 20:581–596
Cuthbert AR (2006) The form of cities: political economy and urban design. Blackwell, Oxford
Davis H (1999) The culture of building. Oxford University Press, New York
Ewing R, Cervero R (2010) Travel and the built environment. J Am Plann Assoc 76(3):265–294
Fainstein S (2011) The just city. Cornell University Press, Ithaca
Girouard M (1985) Cities and people: a social and architectural history. Yale University Press, New Haven
Gorringe TJ (2002) A theology of the built environment: justice, empowerment, redemption. Cambridge University Press, Cambridge
Grineski S, Bolin B, Boone C (2007) Criteria air pollution and marginalized populations: environmental inequity in metropolitan Phoenix, Arizona. Social Science Quarterly 88(2):535–554
Harvey D (1973) Social justice and the city. Edward Arnold, London
Harvey D (1989) The condition of postmodernity. Basil Blackwell, Oxford
Hayden D (1980) What would a non-sexist city be like? Speculations on housing, urban design, and human work. Signs 5(3):170–187
Jacobs J (1961) The death and life of great American cities. Vintage Books, New York
Klosterman R (2011) Planning theory education: a thirty-year review. J Plann Edu Res 31(3):319–331
Knox PL (2010) Cities and design. Routledge, London
Levine MD (1986) Working it out: A community reaction approach to crime prevention. Journal of Community Psychology 14:378–390
Lynch K (1981) Good city form. MIT Press, Cambridge, MA
Michelson WH (1970) Man and his urban environment. Adkinson-Wesley, Reading
Mitchell D (2003) The right to the city: social justice and the fight for public space. Guilford Press, New York
Newman O (1972) Defensible space: crime prevention through urban design. Simon & Schuster, New York
Nyden P, Lukehart J, Maly MT, Peterman W (1998) Chapter 1: neighborhood racial and ethnic diversity in U.S. cities. Cityscape J Policy Dev Res 4(2):1–17
Riger S, LeBailly R, Gordon MT (1981) Community ties and urbanites' fear of crime: an ecological investigation. American Journal of Community Psychology 9(6):653–665
Williamson T (2010) Sprawl, justice, and citizenship: the civic costs of the American way of life. Oxford University Press, New York

Chapter 9
Architecture and Value-Sensitive Design

Jeroen van den Hoven

First we shape our houses and then our houses shape us

(Sir Winston Churchill)

9.1 Introduction: Politics Designed into Artifacts

The philosopher of technology Langdon Winner (1986) has drawn attention to the fact that artifacts can embody values and can be said "to have politics." The case study that he used to vividly drive this point home to the reader concerns the work of the famous New York architect and urban planner Robert Moses. In the 1920s, Moses designed large urban projects in New York. One of the projects that he was involved in was the design and construction of a series of overpasses on New York parkways. Caro's elaborate study of the life and work of Moses (Caro 1974) gives us reason to believe, according to Winner, that Moses designed some of the overpasses intentionally low so that buses taking the poor and (mainly) colored population to the beaches near New York could not drive under them (Winner 1986). Buses in the new design could no longer be routed to the recreational areas. Indirectly, the overpasses thus functioned as a mechanism and barrier separating black and white middle class. Although there is some controversy over whether Moses really intended his design to have the effect of racial segregation, these overpasses provide a clear-cut illustration of the political and morally relevant effects that designs, built structures, and artifacts may have. With his account of "The Politics of Artifacts," Winner was one of the first to point systematically to the value-ladenness of artifacts. According to Winner:

> The things we call "Technologies" are ways of building order in our world. Many technical devices and systems important in everyday life contain possibilities for many different ways

J. van den Hoven (✉)
Faculty of Technology, Policy and Management, Section Philosophy/3TU. Centre for Ethics and Technology, Delft University of Technology, Delft, The Netherlands
e-mail: m.j.vandenhoven@tudelft.nl

C. Basta and S. Moroni (eds.), *Ethics, Design and Planning of the Built Environment*, Urban and Landscape Perspectives 12, DOI 10.1007/978-94-007-5246-7_9,

> of ordering human activity. Consciously or unconsciously, deliberately or inadvertently, societies choose structures for technologies that influence how people are going to work, communicate, travel, consume, and so forth over a very long time. … In that sense technological innovations are similar to legislative acts or political foundings that establish a framework for public order that will endure over many generations. (Winner 1986, pp. 28–29)

There are many examples of architecture and civil engineering that Winner could have chosen to illustrate his claim "that artifacts have politics." The boulevards in Paris designed by Haussmann in the mid-nineteenth century express a certain grandeur and provided ample space to "flaneurs," but they were also convenient channels for military logistics and the rapid dispatching of soldiers and police. Rykwert (2000) draws attention in his book *The Seduction of Place* to the fact that the headquarters of the UN in New York and Geneva, the EU in Brussels, and the Unesco in Paris have all been designed in disharmony with their local urban environment so as to express a modicum of inaccessibility, universality, and impartiality.

The garden cities at the turn of the century that were built across Europe embodied the ideal that low-income families were also entitled to a green environment – in stark contrast to the unhealthy slums of the big cities of the late nineteenth century. A number of ideals were built into these designs: social cohesion, privacy, responsibility, solidarity, and hygiene.

Not only important ideas and central values may be embedded and expressed in the built environment, also everyday and mundane normative considerations may be shaped with constraints and affordances for the actions and thoughts of users. Anyone who has been to IKEA knows that visitors to IKEA stores are forced to progress through a carefully designed maze that has as its sole purpose to increase sales. It is one of the components of their business success. Fast-food restaurants use hard and uncomfortable chairs to encourage people to move on after they have finished their meal, the arrangement of seats at waiting areas of airports is not supposed to be conducive to conversation, and the numerals and displays in elevators are often positioned overhead so that people can look up and do not have to look each other in the eyes, which can be experienced as somewhat awkward in an elevator. The benches in parks may have an arm rest in the middle in order to prevent homeless to sleep on them. In Amsterdam, window sills of new houses at the beginning of the twentieth century were designed higher than in the old neighborhoods of the city so that people could not easily lean out. In this way the architects hoped to prevent the behavior of working class women who used to hang out of their windows. In the same houses the electricity plugs were not all located in the same places so that people would not all end up with the same arrangements of their dining tables and furniture.

Shah and Kesan (2007) provide many examples in their paper "How Architecture Regulates" of the normative function of architecture. According to Shah and Kesan, architecture can (i) communicate and express cultural values and have symbolic meaning. Banks are large, robust, and often made of marble and convey to the visitor that this institution is made to last. It expresses trust, reliability, and security. Good schools typically radiate accessibility and cleanliness. Furthermore,

(ii) architecture may constrain and facilitate certain types of behavior. New Urbanism, for example, has drawn upon social science research which has shown that design can either contribute or distract from civic engagement and thus may or may not lead to better quality of community life. Shah and Kesan provide more interesting examples of how architecture may shape interactions. Irwin Altman has studied design for privacy in the built environment. He concludes that privacy is not by default about isolation and seclusion but about the ability to control one's exposure to others. Design should therefore accommodate the control and freedom to choose the level of accessibility, instead of providing everyone with a separate space. Finally, (iii) our values and moral ideas may get expressed and embedded in the built environment. Fire safety has been built into houses and requirements, since the early Middle Ages, ideas about the presence of women in public building in the eighteenth and nineteenth centuries have led to a relatively small number of toilets for women, because they were previously outnumbered by men in prominent public buildings. Laws about equal accessibility of building to handicapped have led to better accessibility. Ideas of accessibility have been designed in.

9.2 Value-Sensitive Design

All the examples given above are examples of incorporating or embedding particular values and world views in the built environment. This central idea of expressing and embedding values in artifacts is the subject of study that started in computer science and is now usually referred to as *value-sensitive design (VSD)*. One can design for inclusion, for privacy, for trust, for accountability, and for sustainability, in computer systems. In value-sensitive design, the focus is on incorporating moral values into the design of technical artifacts and systems by looking at design from an ethical perspective. It is concerned with the way our acting in accordance with moral values (e.g., freedom, equality, trust, autonomy, privacy, and justice) is facilitated or constrained by technology (Friedman 1997; Friedman and Freier 2005). Value-sensitive design focuses *primarily* and *specifically* on values and requirements of *moral* import. Other frameworks tend to focus more on functional requirements such as speed, efficiency, storage capacity, and usability. Although building a user-friendly technology might have the side effect of increasing a user's trust or sense of autonomy and freedom, in value-sensitive design, the incorporation of moral values into the design is a primary goal instead of a by-product. Value-sensitive design is at the same time, as I have argued (Van den Hoven 2005), a way of doing ethics that aims at making moral values part of technological design, research, and development. VSD can only be used if we manage to be explicit about the variety of moral reasons for desirable features of systems and can formulate them as "non-functional requirements" and have a transparent way of decomposing them into more detailed functional requirements. If we cannot do this, we will only replace the obscure by something which is more obscure.

VSD helps us to look more specifically at ways of reconciling different and opposing values in engineering design or innovations (Van den Hoven and Weckert 2008). This idea (see Van den Hoven et al. 2012a) can be illustrated as follows. As a society, we value privacy, but at the same time we value security and the availability of information about citizens. The pursuit of these values creates a tension which is exemplified in the debates about ubiquity of closed-circuit TV (CCTV) cameras in public places. We either hang cameras everywhere, and thereby create the desired level of security in that area, but give up on our privacy, or we respect privacy and refuse to hang cameras everywhere but settle for less security. Ideally we want both privacy *and* security. Smart camera systems may allow us to have our cake and eat it, in the sense that their smart architectures may allow us to enjoy the functionality the technology can offer and at the same time respect the moral constraints on the flow and availability of personal data that privacy requires. The police may use software tools to prevent operators of CCTV cameras to look inside houses. The smart technology underlying the relevant innovations allows us to configure the system in such a fine-grained manner that the systems allows one to use the advantages and functionality the technology offers without actually compromising data protection norms. Instead of an all-or-nothing matter, smart privacy-enhancing technology (this is also called "privacy by design") may allow us to stipulate who gets access to which recordings, on which conditions, how long the images are stored, and how they may be used and merged with other databases. Innovations of "smart" technologies often manage to reconcile previously irreconcilable values or preferences by design.

I have dubbed this notable shift in perspective in moral matters "The Design Turn in Applied Ethics" (Van den Hoven and Weckert 2008; Van den Hoven et al. 2012). The basic idea here is that design is a respectable ethical category. Instead of taking human character or a person's actions as the unit of analysis and the object of moral evaluation, it seems sometimes highly relevant to be able to ask questions about the moral quality of a *design*. We need to be able to evaluate proposals to change the world and undertake this evaluation from the point of view of moral values. In the last decades, the work of John Rawls gave rise to talk about design in ethics. Thinking about social justice can, in the context of Rawls' theory, be described as formulating and justifying the principles of justice in accordance with which we should design the basic institutions in society. Thomas Pogge, Russell Hardin, Cass Sunstein, Robert Goodin, Dennis Thompson, and others (Van den Hoven and Weckert 2008) have taken moral theory and applied ethics a step further down this path of semantic descent and practicality. Not only do they want to offer applied ethical analyses, they also want to think about the economic incentive structures and technological conditions and institutional and legal frameworks that need to be realized, if our applied analyses are to stand a chance in their implementation and contribute to bringing about real and desirable moral changes in the real world. Design in the work of these authors is primarily focused on institutional design, but the design turn clearly brings into view the design of socio-technical systems, technological artifacts, urban planning, and architecture.

9.3 Design Against Crime

An example in urban planning and architecture which forms a clear exemplification of the design for values perspective is design for security or design against crime. Human safety and security is now no longer seen as an add-on but construed as a value that needs to be accommodated in design and in the early stages of planning. No amount of policing will be able to deal with crime in public spaces that provide systematic and structural opportunities to thieves and criminals. Medieval castles and fortifications are examples of what is called "target hardening" to reduce vulnerability to attacks and invasion by the enemy. Another historical example is provided by the eighteenth-century philosopher Jeremy Bentham who thought about the ideal prison and exclaimed in his treatise on the subject: "Morals reformed — health preserved — industry invigorated — instruction diffused — public burthens lightened — Economy seated, as it were, upon a rock — the gordian knot of the poor-law not cut, but untied — all by a simple idea in Architecture!" The idea here is that security and control over prisoners is greatly enhanced by the design of a dome-shaped prison with a guard in the middle who can oversee everything. Essential here is that the inmates think that there is someone on guard. The design was appropriately named "Panopticon." The Bentham idea has come round in recent times in a field of research and architecture and planning referred to as design against crime, or "crime prevention through environmental design" (CPTED) (Poyner 1983). We will have a closer look at this evolving field to illustrate how VSD ideas may work in architecture and urban planning.

In an overview article (Katyal 2002), "Architecture as Crime Control," the author embraces the design for values perspective and applies it to security. The author correctly associates the focus on values for design with its origin in IT. The author quotes Larry Lessig who was one of the first to make the idea popular that software is a regulatory force in the twenty-first century, and that IT architectures in important ways help to shape and constrain our lives, actions, and experiences. The same applies in architecture and urban planning: "Some Architects have outlined mechanisms for crime prevention through principles of design" (Katyal 2002, p. 1048). "Architects influence in subtle ways the paths by which we live and think" (Katyal 2002, p. 1048). "The law uses architecture as an expressive tool to embody certain commitments" (Katyal 2002, p. 1048).

Jane Jacobs and Oscar Newman in the 1950s started to criticize the architecture of their time because of what they considered the withering away of the public sphere. This atrophy of community and social dimension of the public spaces made it vulnerable to crime and criminal behavior. Their writings formed an important inspiration to thinking about "design against crime."

Architects cannot but start from a social and moral requirement of well-being of future users. How could they start on the basis of some other value? Well-being however is a broad and vague value. It will only get a definite and meaningful content if it is decomposed in value components such as safety, security, health, freedom, and dignity. Safety, health, and security are fundamental and are preconditions for pretty much everything else in a human life that is worthwhile wanting. So the task for the architect seems straightforward, but what does it mean to build a *secure* environment? Different people may have quite different conceptions of security. Jane Jacobs in *The Death of the Modern City* focused on public space and the sense of community and social control that is associated with it, that is, *natural surveillance*. Others focused on *territoriality* and a sense of *property* and *responsibility* that typically goes with it. Some introduced the notion of *defensible space*.

Jacobs decomposed design for security in four subrequirements: (i) natural surveillability, (ii) territoriality, (iii) community building, and (iv) protection of targets of crime. This is markedly different from views that focus solely on "target hardening" and which take a much more technical approach. Target hardening points in the direction of better locks and thicker walls. Natural surveillability in turn is decomposed in three types of mechanisms: (i) diversity of building, (ii) building design, and (iii) lightening. Lightening, for example, can be decomposed in (i) intensity of lightening (empirical studies indicate that crime is more likely to take place under the five-lux level) and (ii) the spread and homogeneity of the lightening and the color of lightening. The often used yellow light, for example, is associated with crime, spookiness, and lack of security. An Illumination Engineering Society is dedicated to this area of research.

We can see how an initial value commitment of architects to well-being of future users – as a nonfunctional requirement – of their designs can be decomposed in natural surveillability, lightening, and the details of illumination technology.

9.4 Conclusions

Design is in important ways about moral values, and the study of moral values in the twenty-first century cannot do without design. The recent interest on both ethics and design for the ethics of design has given rise to a "design turn in ethics." Value-sensitive design or design for values is a growing field of research. We have illustrated how this manifests itself in architecture by means of a discussion of design for security or design against crime. We have seen how the concern in architecture for the well-being of future users of architecture and built environments takes us all the way to discussions of street lamps and lanterns. Architects and urban planners of the future need to be able to shuttle back and forth between lofty moral ideals and fundamental values and the details of their designs. They will have to be ready to defend the details of their designs in terms of their ideals and defend their ideals in light of the details of their design.

References

Caro RA (1974) The power broker: Robert Moses and the fall of New York. Knopf, New York

Friedman B (ed) (1997) Human values and the design of computer technology. Cambridge University Press/CSLI, Stanford University, New York

Friedman B, Freier NG (2005) Value sensitive design. In: Fisher KE, Erdelez S, McKechnie EF (eds) Theories of information behavior: a researcher's guide. Information Today, Medford, pp 368–372

Katyal NK (2002) Architecture as crime control. Yale Law J, 1039–1137

Poyner B (1983) Design against crime, beyond defensible space. Butterworths, Boston

Rykwert J (2000) The seduction of place. Oxford University Press, London

Shah RC, Kesan JP (2007) How architecture regulates. J Archit Plann Res 24(4):350–359

Van den Hoven MJ (2005) Design for values and values for design. Information age +. J Aust Comput Soc 7(2):4–7

Van den Hoven MJ, Weckert J (2008) Information technology and moral philosophy. Cambridge University Press, Cambridge

Van den Hoven MJ, Lokhorst GJ, Van de Poel I (2012a) Engineering and the problem of moral overload. Sci Eng Ethics 18(1):143–155

Van den Hoven MJ, Miller S, Pogge T (2012) The design turn in applied ethics. Cambridge University Press (Forthcoming)

Winner L (1986) The whale and the reactor. A search of limits in an age of high technology. University of Chicago Press, Chicago

Chapter 10
Designing for Meaning: The Designer's Ethical Responsibility

Stanley M. Stein and Thomas L. Harper

10.1 Introduction: The Crisis in Contemporary Culture[1]

Our contemporary "first" world societies seem to be drifting in a state of cultural crisis. This has been notable for the past several decades. As planning theorist John Friedmann (1993, p. 482) put it sometime ago:

> What we are living through in the final decades of this [20th] century is something altogether different. It is nothing less than the collapse of the Euclidean world order of stable entities and common sense assumptions that have governed our understanding of the world for the past two hundred years.

Rather than abating, this crisis seems to have become chronic and perennial, though often ignored. It relates to profound changes in how we see the world (our conceptual frameworks or paradigms), in how we come to know (epistemology), in how we decide what we ought to do (morality or normative ethics), and in how we find meaning in our lives.

The first wave of change came from the modernist replacement of religious faith by science as foundational source of knowledge and justification. This led to scientism – the claim that the scientific method was the only source of knowledge – and the dominance of a mechanistic and instrumental mode of thinking. The second wave[1] was the postmodernist questioning of the very possibility of any sure foundation for knowledge, leading to a loss of the modernistic faith in science (Harper and Stein 2006). The result of this challenge was an erroneous[2] (but widespread) view that there is no longer any way to justify our beliefs and values. Our contemporary (economically) advanced societies seem to be under the sway of a confused combination of modernist "instrumental reason" and postmodernist "soft relativism," leading to a narrow and self-absorbed search for "authentic identity" and a loss of vigor in political culture (Taylor 1991).

S.M. Stein • T.L. Harper (✉)
Faculty of Environmental Design, University of Calgary, Calgary, AB, Canada
e-mail: stein@ucalgary.ca; harper@ucalgary.ca

C. Basta and S. Moroni (eds.), *Ethics, Design and Planning of the Built Environment*, Urban and Landscape Perspectives 12, DOI 10.1007/978-94-007-5246-7_10,

These changes were accompanied by dramatic shifts in the way objects are designed, made, sold, and consumed, which in turn has altered the way we live our lives. Handmade, locally, regionally, or nationally produced objects which were connected to our "place" once added meaning to our lives. They have been largely replaced by mass-produced machine-made objects designed for global markets. Disconnected from place or culture, they no longer contribute any significant meaning to our lives.

Our technologically sophisticated "first" world which has eventuated fails to provide artifacts, buildings, and environments which facilitate human flourishing. We believe that the design professions (e.g., industrial/product design, architecture, urban planning/design, regional/environmental planning) have played a significant role in creating or exacerbating this crisis of meaning; the results of their work have contributed significantly to alienation. While designers cannot resolve this crisis alone, we believe that they do have the potential to add meaning to our artifacts and environments and the ethical responsibility to do so.

The first section (10.2) traces the alienation and loss of meaning in our artifacts and environments, with changes in the technology of production and the ascendancy of the mechanistic world view. The second section (10.3) traces the roots of the instrumentalist, scientistic world view to the scientific aspect of the Enlightenment. The third (10.4) presents the humanistic view, also rooted in the Enlightenment, with its emphasis on the authentic experience of the autonomous person. It explains how this view is not in conflict with the scientific view, and advocates a new emphasis on designing and planning for meaning. The fourth and final section (10.5) gives some ideas regarding designing and planning for meaning at different scales, ranging from the house to the natural environment.

10.2 Contemporary Culture and Alienation

The meaning of a designed object or environment to its user can be considered on a *continuum.* At one end is an object handcrafted by its user or designer. In 1800, most consumer products – vehicles (horse-drawn), tools, hardware, houses, furniture, furnishings, clothes, and linens – were made locally or regionally or within the consumer's own country. Particularly outside large cities, many products (including homes) were handmade by the user, or by someone known to them. At the other end of the continuum is an object designed and produced in a completely mechanized process (as described below).

10.2.1 The Mechanistic Shift

A great shift began with the introduction of the factory c.1800 and the development of the industrial city by c.1850. (This shift came later in North America.) In the early 1900s, the shift was accelerated by the mechanized assembly line. Over the next 50–60 years, production was progressively more mechanized. Products were still designed by persons but more and more were mass produced. An increasing number of products came from further away, as transportation, communication, and refrigeration technologies enabled goods to be shipped over much longer distances.

Over the subsequent 50–60 years, the ever-increasing speed and sophistication of computers introduced, and then broadened the scope of, computerized design and computerized production. Thus, not only production but also the design process itself become more and more mechanized. In addition, information and communication technology, abetted by low transport costs (particularly inexpensive oil), allowed production to be globally dispersed. Today, few of our products come from our locale, region, or country. A large proportion are imported, often assembled from components manufactured in several different countries. Almost *none* of our products are handmade or self-made: handcrafted goods are now high-status luxury goods. Over time, as the chain linking designer and user lengthened, the once-intimate link between them has been weakened or completely sundered.

On our continuum of meaning, the opposite end to handcrafted would be reached when the product is designed by a computer algorithm. Then the user experience is the antithesis of the experience of a handcrafted object. Only slightly less disconnected is a design by a person using a computer. Just as disconnected to those affected are government policies, plans, and decisions made by faceless functionaries following rigid sets of rules, which is more common as organizations become larger, more bureaucratized, and more mechanistic (Hummel 2008).

10.2.2 Mass Production and Consumerism[3]

Fashionable mass-produced consumer products are slick, sleek, shiny, bright, and perfect. But the illusory nature of their perfection rapidly becomes apparent as their colors and styles go out of fashion. Clothing (particularly "fashion") is probably the epitome of transient perfection. Close behind are electronic products and appliances – computers and peripherals, cell phones and tablets, TVs, stereos, video and music players, radios, and household appliances. These products offer no opportunity for users to feel any sense of participation in their creation, nor to experience any authentic identification with them.

With mass production, the maximization of shareholder profit requires large customer bases for the same (or similar) product sold in world markets; they generally cannot have any qualities connecting them to place or culture. While they have extrinsic value because they are useful, and perhaps in style, these products have no *intrinsic meaning* to us.

Within this production system, designers are far removed, both physically and culturally, from users. Their designs reflect neither cultural nor local differences (Badke and Walker 2007). Product designers are largely oblivious to the dehumanizing effects their designs have on users. In any event, when roles in the design, production, and marketing process are specialized and divided, no one feels responsible. Each person thinks "It's not *my job* to give the consumer a sense of meaning."

Manufactured products are increasingly disposable; so-called consumer durables become technologically or stylistically obsolete even before they wear out. Few are cost-effective to repair. When they cease to function, we replace them. And we have little compunction about discarding them, because the product *is* out of style and we don't want to be seen as "dated." These types of products are the ones we feel no authentic relation to and dispose of most readily. Although recycling has improved, landfills are filled with these objects which objectify us.

10.2.3 Alienation and Dehumanization

Badke and Walker (2007) assert that

> Western societies are hooked on consumption, and this exhibits similar traits to other addictions such as alcohol and nicotine. As consumers, we seem never to be satisfied. We make a purchase and "get our hit", but the thrill soon wears off, and we return, again and again, to consumer more… we are constantly told that the latest product will bring us satisfaction, happiness, and fulfillment. (*ibid*)

They argue that both consumers and designers need an examination of their values and lifestyle akin to 12-step programs like Alcoholics Anonymous.

In general, the search for meaning has been preempted by a search for efficiency – for "the one best way" (Ellul 1967). This reification of "efficiency" has altered the way we see ourselves, our relationships, our artifacts, and our environments. The resulting products and environments do not contribute to meaningful and authentic lives, lives where we feel in control. When a mechanistic, reductionistic approach is taken to design or planning, the likely outcome is an environment that is alienating. When a person is surrounded by such environments, "finding himself [sic] nowhere outside himself, he can find himself nowhere within" (Scruton 1979, p. 245). Users/occupants are in danger of becoming alienated persons whose "activity is that of a body in the grip of a machine, not of a rational agent acting out of a sense of value. In his [sic] own eyes he is what he conceives himself to be in the eyes of the world – a means, not an end, an organism, not a man" (*ibid*).

Once we might have thought of ourselves as being in control of our physical world, using products and inhabiting buildings and environments which express our values. Now we are in danger of losing our full humanity, becoming objects expressing the mechanistic worldview of modernism/scientism.

Design, according to styles, or fads, or abstract universal principles (e.g., modernism or other "universal" design approaches, single-use zoning, sustained yield), does not establish any connection with users. The modernist International Style deliberately sought to divorce the building from any connection to its context – physical or cultural. The ideal of modernism was a building which functioned like a machine; Le Corbusier (1923) famously declared that "a house is a machine for living."

Designers should ask themselves: Is a life lived inside a machine a meaningful life? Or is it more like the nightmare that Chaplin visualized literally in the movie "Modern Times" when his character is enmeshed in the gears of a giant machine? Most people desire experiences that are genuine/authentic (the opposite of Nozick's experience machine discussed in 10.4.6).

In the early 1950s, Levittown (the first "mass-produced" or "corporate" suburb) extended this lack of meaning and identity to the family home in order to minimize the production cost and thus increase the potential market. This new technology rapidly spread across North America during the 1950s. Along with other factors (government mortgage guarantees, government-funded urban highways, cheap gasoline), it widened the availability of the "American Dream."

10.2.4 Trivial Identity and Meaning

Many North Americans do not seem to identify strongly with categories which carried significant meaning to previous generations (e.g., religious affiliation, race, ethnicity, service club, bowling league). Instead, they often attach greater meaning to their physical environments, and the ownership and consumption of artifacts and products. This increases the impact of design on their lives.

A major goal of *marketing* is to exploit the need for meaning by manipulating potential buyers into feeling a superficial sense of meaning and identity from their purchase of products. Increasingly, we derive our identity from the products we buy, the spaces we inhabit, and the activities in which we participate (Sparke 2006). Advertising invites buyers to acquire an identity by consuming a *brand name* product: wearing Calvin Klein or Nike, driving a BMW or a Mustang, drinking a whiskey that is "handcrafted," etc. And we are invited to share the supposed taste of our favorite athlete or celebrity by using a product endorsed by them. Motor vehicles were among the earliest sources of ersatz identity. In the 1950s, North American males often viewed themselves as "Ford men" or "Chevy men." Each vehicle's advertising strove mightily to convince buyers (who were then mostly male) that their car would attract the most beautiful woman.

Marketing also attempts to lure consumers into competitive consumption, in which products are purchased in order to position them favorably in comparison to others in their social group (Lansley 1994). Marketing often suggests that certain products are associated with an exclusive lifestyle – purchasing them expresses good taste and aligns you with an elite lifestyle. This is rather ironic, considering the necessity of blandness for mass-produced products, but the "higher-end" products are distinguished by additional features and an appearance of higher quality.

The idea of consumerism in general is promoted by the concept that each of life's problems can be solved by purchasing the appropriate product. In fact our existence would be satisfying and complete "if only we bought the right things" (Oskamp 2001). Another irony is that fashion cycles are intended to ensure that any such satisfaction will be transitory by creating *dissatisfaction* with the things we already own.

Charles Taylor views the satisfaction of self-absorbed forms of individual expression by the "consumption of quick, shoddy, replaceable commodities" (Taylor 1991, p. 6) as a "loss or decline, even as our civilization develops" (*ibid*, p.1). Our conceptions of progress and innovation become shallow and unreflective. Our modernist technical focus has made us expert at doing things well but uncertain about why we are doing them. Perhaps we should stop and reflect on questions like "What is the point?" Are we adding any real value to our lives with all our consumption?

In order to understand what is happening here, we need to look at where our culture has been, why it is in a perennial crisis, what are the factors contributing to the crisis, where it might be headed, and what is the designers' responsibility in this situation.

10.3 The Instrumental and Mechanistic Worldview

10.3.1 The Origins of the Crisis[4]

The various professions which practice design arose in the nineteenth and early twentieth centuries from the development of modernist ideas in the eighteenth century. The "Enlightenment project" challenged the old premodern order, where religion provided an absolute foundation for certain knowledge (truth) and justification of all kinds – empirical, moral, aesthetic, and religious. This single foundation gave the premodern worldview unity and coherence. Everything and everyone had a fixed place (defined by birth) within the divine order: "This hierarchical order in the universe was reflected in the hierarchies of human society…at the same time as they restricted us, these orders gave meaning to the world and to the activities of social life" (Taylor 1991, p. 3).

For premodern persons,

> the great cathedral of Notre-Dame de Paris was…not simply an object, a pile of stone and glass artfully arranged. It…was an extension of their collective existence as humans, or perhaps a projection of their being, a reaching up to God by man and a simultaneous reaching down to man by God, as in Michelangelo's *Creation of Adam* in the Sistine Chapel. Men and women wore it…like a cloak…and it exalted their existence, in ways most of us can no longer even imagine. (Rowland 1999, p. 59)

Their shared worldview gave a unified *meaning* to their existence, their experience, their artifacts, and their world.

By the end of the seventeenth century, skepticism and the questioning of tradition, custom, and authority were shaking this foundation, with the claim that science could provide an alternative unshakable foundation for knowledge. This augmented the effects of the sixteenth-century Protestant Reformation, which challenged the doctrines, rituals, and ecclesiastical power of the Roman Catholic Church and its claim to be the only path to God. By the end of the eighteenth century, advances in scientific knowledge were being applied to technology (leading to the

"Industrial Revolution"). The entire social, political, and religious order was challenged by radical ideas, disseminated by a new technology – the printing press. The unity of the premodern view became more and more eroded.

Two key aspects of modernism[5] led us to the contemporary situation: science and humanism. These two aspects point in directions which are often seen as being in conflict.

10.3.2 Science and Scientism

Science replaced religion as the foundation of knowledge about the world. Science claimed to produce an *objective understanding* through the observation of regularities. The *scientific method* provided a procedure for determining which regularities are *causal*. This method specifies a sequence of activities[6] that *legitimize* or establish empirical knowledge. Scientific explanations were very successful in serving human purposes, by enabling the manipulation and (apparent) control of the physical world.

The success of science led many modernists to inappropriately expand what they believed the role of the scientific method should be. They proclaimed the scientific method to be the *only* method for determining what is true. Their claim that all claims to knowledge (religious, social, moral, aesthetic) must be translated and *reduced* to hypotheses that can be tested using this method. Language that cannot be translated into scientific language (so that its claims can be falsified empirically) is held to be meaningless. Habermas (1984) has called this misapplication of the scientific method "the fallacy of scientism": the claim that science and its method is the *only* source of knowledge.

Adherents to scientism applied the Newtonian idea of the "clockwork universe" (Dolnick 2010) to the analysis of society, seeking to develop "metanarratives" – universal theories that explain social reality in a universalizable and deterministic way.[7] The study of persons, their interactions, and their societies became known as "social science." Over time, ordinary people (nonexperts) started to accept the reductionistic *assumptions* made by social sciences as accurate descriptions of reality.

Over more time, many people come to view the assumptions as *normative*. For example, empirical economics assumes the "economic man," one who maximizes his own self-interest, generally (long-term net) happiness, by the consumption of goods and services which satisfy his desires. Initially, no claim was made that this was an accurate or holistic description. Now it is taken as normative in several ways. One is the use of benefit-cost analysis to make government decisions. The other is more pernicious: the critique of behavior which does not fit the assumption. If your actions do not fit this "model," there is something wrong with you. You are "irrational." This subtle slide from the descriptive into the normative is often completely unrecognized.

10.3.3 Scientism and Design Professions

One result of applying scientism to practical decision making is a sharp distinction between fact and value and between means and ends. Rationality is limited to *instrumental* rationality: finding the best means to given ends. This narrow stipulative definition of rationality has had a pervasive impact on all realms of life:

> Ethically, modernism is *utilitarian* – the most scientistic approach to morality, that reduces it to a cost-benefit calculation… Politically, the modernist form is *representative democracy*, with its sharp bifurcation of politics and administration. Organizationally, the modernist form is *bureaucracy,* with its emphasis on hierarchical structure, routine and instrumental rationality. All of these taken together form a fairly consistent and coherent world-view… This world-view still dominates a good deal of institutional planning practice, and is increasingly influential in the "third world" as various "first-world" (or western or northern) agencies assist it to "modernize" at a frenetic rate. (Harper and Stein 2006, p. 5)

Professionals arose as the *experts* who used the best scientific knowledge (not available to the ordinary person) to select the best means, without regard to the ends. Because rationality (evaluation or critique) of ends or values is ruled out, professional design education has been primarily a matter of conveying *technical expertise* (constrained by codes of professional ethics).

10.4 The Humanistic View

In order to understand the potential for designers to contribute to meaning, we need to understand the importance of artifacts, buildings, and environments to the authentic meaning of persons' lives. This requires an understanding of the humanistic view, which is in sharp contrast to the scientistic view. The "liberal" or humanistic aspect of the Enlightenment can be traced back to the ideas of John Locke, Immanuel Kant, and Thomas Jefferson. The core belief is that the *autonomous* individual person is the source of value and the appropriate object of moral (ethical) and political concern. This is the aspect of modernism that society and the design professions lost with the growing dominance of instrumental scientism. We believe it is worth preserving, and that designers and planners have an ethical responsibility to assert its importance against the dominance of the mechanistic view.

10.4.1 Particular Event, Many Descriptions

The scientistic view creates much confusion when it tries to *reduce* all accounts of a particular event to a scientific one. For example, thinking is "nothing but" brain function, and thus displaces humanistic accounts of intentions and choices. Descriptions are general, not particular. There can be a variety of different, consistent descriptions of a particular event. Thus, we may provide a set of descriptions from what we will call a humanistic perspective (the perspective of a "person" as defined below). These may be called "social constructions," if you like. There is a common scientistic misconception that a social construction cannot have an independent existence, nor be real, nor authentic. We argue that this is not true. The existence of nature or wilderness, for example, is real and authentic, even though its existence depends on the character of certain relations with persons. An object or an environment may have both a character in relation to persons and an independent physical existence. Recognizing the truth of both descriptions in no way denies their identity nor their reality.

A particular *event* can be legitimately *described* in *many* different ways (using different conceptual frameworks). For example, humans can be described in a humanistic way as persons (with intentions) or in one of many scientific ways as animals (behaviorally or physiologically) or as collections of molecules, atoms, particles, etc. One account (e.g., intentional) is not reducible to another (e.g., behavioral). Nor is one account (e.g., scientific) superior to (more objective, more real than) others. The *choice* of appropriate account is contingent on purpose or *interest.* Why are we describing the human being(s)? What is the problem we want to address? What kind of intervention are we contemplating? The question of which account is appropriate depends on how well it serves our interest or purpose. The question is *not* how well the concepts represent, or correspond to, the "real world" (Rorty 1981).

A shift to a new description is *not* the result of induction or empirical generalization. It requires a *conceptual* shift (a new conceptual framework). So, my arm going up (a particular event, under a particular description) can be described in a causal way as a physiological activity. Or it can be viewed as supporting the election of a president, because my arm going up can be described as "a vote" using the *conceptual frame*, not of physiology, but of political democracy, which is humanistic, intentional, and relational. If your interest is in the election, then you should think of the behavior as that of a being having intentions, beliefs, hopes, and ideas, that is, a person who is fully human (in the sense just discussed).

Other ways of describing this key distinction are internal vs. external, or intrinsic vs. extrinsic, or meaningful vs. empirical, or intentional vs. causal. A relation is *internal/intrinsic* if that relation requires an intellectual act. That is, we must *know or understand* that with which we have an internal relation (Scruton 1979). This kind of relation is *meaningful, value laden.* The relation is *not* merely an *empirical* one, which can be captured by *causal* explanation. When a relation is merely *causal/extrinsic,* that relation is one between things: object and object. When a relation is *intrinsic,* it is one between subject and subject or between subject and *meaningful object* (i.e., meaningful to the subject).

Again, this is not to deny that there are two-way causal relations between ourselves (as members of the species Homo sapiens) and the natural environment. We affect the natural world and it affects us. We emit particulates and gases into the atmosphere, which in turn cause innumerable effects on our lives, such as smog, acid rain, and global warming. These impacts are appropriately described in the causal language of science (biology and ecology). But our relation to that natural world is more than causal; it is also *meaningful* and, as such, is *value laden,* including moral and aesthetic (and for some, religious and spiritual). It involves how we think about, or conceive of, our environments and objects in them (Stein et al. 1999) and has implications for the way we treat them. Furthermore, internal relations are *normative,* in that they imply *standards*; we appeal to these standards in order both to understand that to which we relate and to evaluate it. Thus, we are using a normative framework if we describe our emissions as "pollution" or their effects as "negative." The term *authenticity* is used to evaluate the nature of this kind of a relation (10.4.5). To be authentic in our sense, a relation must be (at least in part) internal (*ibid*).

10.4.2 The Person

According to Kant (1785), the value of the individual person is a "transcendental deduction" from the fact that knowledge requires the ability to (i) make *judgments,* via the application of disparate concepts to experience (empirical reality) and (ii) *reflect* on the object (our self) which engages in the process of judgment making.

This ability, to reflect on our own process of judgment, leads to self-awareness – knowledge of our "self" as distinct from all else and as an enduring entity through time. This self is not the object of perception, but the *subject* of perception, judgment, and action. This awareness of the subjective self allows us to (i) form a concept of ourselves through time as something of *value*; (ii) generate a long-term, enduring *concept* of our *own life* as *meaningful*; and (iii) formulate *plans* which are intended to implement this concept of our own life.

A necessary condition of formulating this concept is that the self must endure through time. *Memory* of the past and some sense of the future are essential. A meaningful self-concept is rooted in the past and extends into the future. It is radically different from an empiricist concept of the self that behaves on impulse or in reaction to its environment. The reason that persons are the source of value and the objects of moral concern is that a choice of a means to achieve a goal *matters*; it makes a difference. The idea of a choice being valued is intelligible only if the choice is consistent with our own self-concept; the choice must be reflective and critical. Thus, the Kantian conception of a person is a being[8] which (i) is thinking, aware, and *self-conscious* over time; (ii) *intentionally* formulates goals and acts to attain them; (iii) is capable of appreciating the *attainment* of goals; and (iv) is capable of experiencing *happiness* and *suffering.*

10.4.3 The Autonomous Person

This valuing of individual autonomy relates to the core underlying moral belief that each person matters and *matters equally.* Kant expressed this in his dictum: each person should be treated "never simply as a means, but always at the same time as an end" (*ibid*). This ethical duty places (negative) side constraints on our behavior, that is, we should not interfere with another person's pursuit of their own goals, without a justification for this interference. These side constraints express the inviolability of *each* person: "there is no justified sacrifice of some of us for others…" (Nozick 1974, p. 33). There is no moral balancing act which can weigh one individual's worth against another's. In other words, no appeal to utility or social good[9] or the "public interest" can justify using any individual as a *mere means* to our own ends. Each of us leads separate lives. We each want more than just experiences and emotions; we want to do things within the context of a conception of life that has *meaning* to us. Thus, value is placed on the autonomous individual person: one who is free, rational and reasonable, capable of making choices, of formulating a conception of a good and meaningful life, and of critically evaluating and modifying this concept.[10] Such a person is not just free, but free to pursue what they decide is *worth doing* – free to lead a worthwhile, *meaningful* life.

10.4.4 Identity

Our self-concept forms the core of our identity. But it cannot be formed in isolation. To know your "self" (to be aware that *you are*), you must understand that there is a world. And you come to understand your relationship to it, as you learn to communicate with other persons about their understanding of the world and their relation to it. Our identity essentially arises out of this relational process. We do not create ourselves out of nothing. The person we become presupposes a social framework and a world, in a process that Donald Davidson calls triangulation.

The argument for *triangulation* is a logical one. As Donald Davidson points out, "the ultimate source of both objectivity and communication is the triangle that, by relating the speaker the community and the world, determines the content of thought and speech" (Rorty 2000, p.15).

The core notion is that our concept of "self" arises simultaneously with our concepts of others (persons) and of the world. And this relation arises simultaneously with our ability to communicate (i.e., with our learning a language). In order to have an understanding of the world and to know when we are right, we need another person to correct us when we are wrong. It follows that my ideas of (i) who I am and (ii) of being right, requires a world and an understanding of it; and in order to understand it, I need a relation with another person. Then I have concepts of (i) myself, (ii) other selves, and (iii) the world. These are *necessary conditions* for literal meaning (Fig. 10.1).

This is an inherently *dialogical* process: "…the selves that arise out of that process are dialogical all the way down…there is no private core on which to build …" (*ibid*, p.16). And the way to evaluate our thoughts and actions (with regard to other persons, to objects, and to our natural and built environments) is *relational.* As Taylor argues, a fully human life has a "fundamentally dialogical character. We become full human agents, capable of understanding ourselves, and hence of

Fig. 10.1 Triangulation

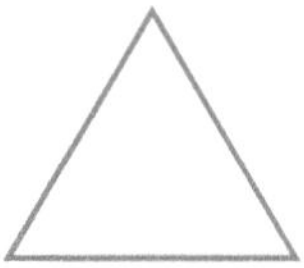

defining an identity, through our rich languages of [all forms of] human expression" (Taylor 1991, p. 33).

These relations are dynamic. Each person's identity is strongly influenced by their relationships and the meaning that these relations have for them. Their relation to each of their environments is shaped by their conception of that environment, which is a social construction. As already discussed, this is not a denial of physical causality. Nor does it mean that the physical objects in the environment would not exist without us, but that what gives them *meaning* to us is our conception of them.

10.4.5 Authenticity

Our relations with each other, as well as the world, have a normative character, a character that is communicated through the use of "thick concepts" (Murdoch 1970; Williams 1985). These are concepts that are descriptive but also have a component that express our normative and evaluative interests. In other words, they are essentially *value laden.* Examples are concepts such as "wilderness," "nature," "environmental," "urbane," "civil," and "professional." Thick concepts express qualities and relationships that are not merely causal/external but humanistic/internal. *Authenticity* is a prime example of such a thick concept: it refers to the normative quality of relationships. Our *authentic* identity essentially arises out of a *relational* process of triangulation. In order to assess the authenticity of objects or environment and our relation to them, we need to share our reflections with another person.

This conception of authentic identity contrasts sharply with a view which has developed within modernist thought. In this view, which Taylor (1991, p. 14) calls "the individualism of self-fulfilment," authentic identity is developed by my "self" in isolation, by listening to my own unique inner voice, which tells me "what is really important or of value." The ideal is to be true to myself, which "means being true to my own originality, and that is something only I can articulate and discover. In articulating it, I am also defining myself" (p. 29). This shallow individualism involves "a centering on the self and a concomitant shutting out, or even an unawareness of, the greater issues or concerns that transcend the self…" (p. 14). This self-defined "authenticity" may be used to justify "rejecting our past as irrelevant, or denying the demands of citizenship, or the duties of solidarity or the needs of the environment" (p. 22). Centering on the self "both flattens and narrows our lives, makes them poorer in meaning, and less concerned with others…" (p. 4). It can lead to viewing all our relations as instrumental, treating other people and environments as mere means to satisfaction of our own ends.

It is important to remember that the relations we are discussing are intrinsic and internal, not extrinsic or instrumental. Culture is a means of expressing this meaning morally and aesthetically. If individuals are not active participants in creating meaning within our culture, the culture fails us. When our relations with the world are not authentic, we become passive victims of outside forces, social and environmental. Individuals lose an active role in creating meaning when powerful interests engage in manipulation, deception, or misrepresentation in order to achieve someone else's illegitimate end. When people are manipulated, they become objects, rather than subjects of their lives. When we lack meaningful contact with the other, we lose contact with ourselves. In eschewing authentic relationships, we become alienated people – alienated from ourselves, others, and the world we live in (Stein et al. 1999). We become more machinelike and less human.

10.4.6 Authentic Experience

What makes an experience authentic? For the answer to this question, we turn to an example of the opposite experience: a thought experiment first described by Robert Nozick (1974, p. 35). He imagined a device that could be programmed to simulate any experience we wanted. Once plugged in to the "experience machine," we have only experiences we want – in other words we don't *do* the things we desire, such as write a book, have sex, or be an architect, we just have the *experience* of doing them.

Why not plug in? No effort, no failure, no hardwork, and no awkward moments – only happiness, achievement, excitement, etc. Why do we not choose to plug in? Because, if the experience machine is the source of what we do, the source of what happens to us, we are an object and not a subject. We are not autonomous persons but *objects* that are acted upon and have a series of *causal reactions,* pleasant though they may be. In Nozick's words, "it is a kind of death" (*ibid*). The idea of plugging into to the experience machine is anathema to us, since to do so is to commit a kind of suicide. Remember that the "I" (the self) is *reflective:* able to judge right from wrong and able to *act* accordingly. On the machine, there would be no right or wrong, and there would be no action. We would be without qualities in the machine. Are we wise, clever, nasty? There is no answer to these questions since everything is simulated. A life "led" in the machine has *no meaning.* All relations between us and the machine are causal/external and not internal in the sense discussed above. On the machine there can be no triangulation because there is no "I," nor are there other persons to whom I can relate. The blob in the machine has no authentic existence.

At the time Nozick postulated the experience machine, it seemed like a science fiction fantasy. The possibility it envisions seems more realistic now. The last time we suggested it to a class, several students said they would consider plugging in – a frightening generational change!

10.4.7 Designing for Meaning: An Ethical Responsibility

We have given a rather simple account of our cultural crisis of meaning, worldviews, economic transformation, and alienation. In focusing on factors relevant to our topic, we have omitted many important factors and forces. A number of books have been written examining the effects of some of these factors and forces from different disciplinary perspectives – for example, in architecture Scruton (1979), in planning Friedmann (1987), in education Bloom (1987), in philosophy Taylor (1991), in sociology Putnam (2000), and in environmental ethics Scruton (2011). However, we believe that the design professions (e.g., industrial/product design, architecture, urban planning/design, regional/environmental planning), having played a role in creating or exacerbating this crisis of meaning in our material world, do have a moral responsibility to address it.

The design professions generally recognize a number of ethical duties inherent in the concept of "professional," and most have recently added a duty related to environmental sustainability. What we are proposing is an addition to, not a replacement for, these responsibilities. This *additional* strong ethical element is inherent in the role of the designer. This element involves designing to give meaning to the individual, to other persons, and to their shared world. Ethics is concerned with doing the right thing in our relationships to each other and to our environments. A prime concern of ethics is the autonomous person. It is not that conditions or states of the person (like their well-being or happiness) are not ethically relevant, but that they alone are not sufficient to fulfill our ethical duty. We argue that the most important ethical aspect of design relates not merely to the functional (utilitarian) purpose of the object designed but to its reflection of the individual person – to its contribution to *their autonomy and to their identity*. The objects and environments we design or plan play a role beyond the functional. They have the potential to enhance (or to inhibit) human flourishing, to contribute to (or detract from) the meaning of the users' lives, to humanize (or dehumanize) them, and to increase their alienation or help them feel "at home" in the world. The design process can raise the level of an object to one with intrinsic aesthetic and ethical value.[11] This potential adds another strong ethical dimension to the design process.

This aesthetic and ethical point is analogous to the logical one made previously regarding triangulation. Our connection to the world around us is enriched and made more meaningful by our connections to other persons. When there is a clear and direct connection between the designer and their work, then it is an expression of the designer. The user of the object experiences the connection and thus experiences a connection to the designer. This connection infuses the object with *meaning* for the user. Particularly when the designer and the user are acquainted (or even have a shared cultural context), the object will have meaning to the user and will more likely support and enhance their identity, their personhood, and their humanity. All three sides of the triangle are completed: the designer expresses themselves in the object, the user experiences this expression in the object, and they also come to understand something of the creator in this experience (Fig. 10.2). The goal is to understand our own identity morally and aesthetically and to feel at home in the world.

Fig. 10.2 Triangulation with a designed object

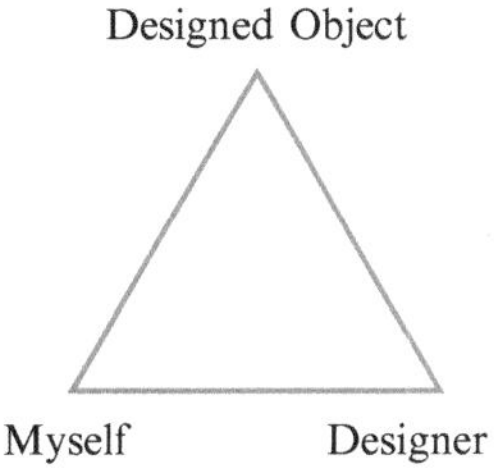

10.4.8 Social Sustainability

We will refer to the creation of conditions which facilitate meaningful lives as *"social sustainability."* The goal is to understand our own identity morally and aesthetically and to feel at home in the world. Recent social/political movements have forced producers of some products to pay attention to environmental sustainability. Caring for the environment requires some empathy with other persons and with the physical world (i.e., a relation to them). Therefore, we believe that "social sustainability," as well as being good in itself, probably happens to be a prerequisite for environmental sustainability. The influence also goes the other way. Many of the ideas touted as environmentally sustainable may in fact be more supportive of *social* sustainability. For example, shopping at farmer's markets or adhering to a "100-mile diet" may be dubious with regard to conserving resources but definitely increases interaction between local producer/sellers and buyers and among like-minded buyers. The result can be a sense of community and added meaning to the buyers' experience.

Something we have created ourselves, or that a loved one has created or given us as a gift, or something associated with our personal history, or with the history of our place has a special (nonmonetary) value for us. Such an object may not be valued for its beautiful appearance, nor for the skill evidenced in its creation, nor for its functional utility. It will be valued for its *inherent* qualities: for its personal or cultural association or for the *expression of its creator* with whom we have some personal experience. Its value and beauty are *inherent* rather than instrumental and *intrinsic* rather than extrinsic. When such an object becomes damaged or ceases to function, we are much more likely to attempt to repair it, rather than simply discarding it. Of course, if we had a role in its design or creation, we would likely be more able to affect a repair because we already have *an understanding* of the object, what it is made from, how it is made, and how it works (Walker 2002).

Our knowledge of an artifact influences our conception of it, our experience of it, and our response to it. For example, we might admire the beauty of a work of art by an artist we admire, but if we discover it is a *fake,* this will *diminish* our view of it. We now see it differently, because a fake is completely inauthentic; it represents deception, a lack of honesty. Its appearance has not changed, but it has lost its intrinsic value for us.

Industrial designer Stuart Walker argues that a material culture that is *meaningful* would, in turn, help alleviate the damaging social (and environmental) consequences of contemporary consumerism. A reframing of our view of physical products and environments is required, together with a creative reengagement with "objects," if we are to find any authentic *meaning* and value in our material world. Walker points to Gandhi's *dhoti* (loin cloth) as an example: "the *dhoti* was much more than a simple article of clothing; the spinning wheel and homespun clothing had social, political, economic and even spiritual importance" (Rühe 2001). "The *dhoti* was a distinct and conscious breaking-away from the 'western' business suit, which he had previously worn during his early career as a young barrister, and as such it was *deeply symbolic*. The *dhoti*...represented self-determination, self-respect, creativity, cultural restoration, independence, and a political and economic statement against colonial rule. Seen in these terms, the *dhoti* [becomes]...a physical embodiment of a philosophy and a set of values" (Walker 2002).

Here again, recognition of this meaning rests on our knowledge about the object, its intrinsic qualities, and the relationship of these qualities to our understanding.

10.5 Designing for Meaning at Different Scales

In this section, we will point to a few ways in which design and planning could enhance the meaning of our buildings and environments.[12]

10.5.1 Housing Design

Except for the wealthy few who can afford the services of an architect, houses in North America are not designed by people who have been educated in design (except perhaps for the technical aspects) or have much conception of the potential for communicating meaning through design.

Contemporary corporate suburbs do offer more choice than they did 50 years ago: buyers may often choose from a number of floor plans, from façade styles like Tudor, Cape Cod, or craftsman and from a wider range of materials and finishes. It is possible, but seems dubious, that the buyer gets a strong sense of personal meaning from these rather superficial choices.

However, when we look at older suburbs, it is clear that many occupants do have a strong interest in adding meaning to their homes by personalizing them through modifications. A tour of "wartime"[13] housing communities (originally very homogeneous in style) in most Canadian cities reveals an amazing array of exterior modifications: covered porches, modified rooflines, additions, a much wider range of finishes, etc. Many innovative adaptations were already noticeable 35 years ago (Galloway 1978). The ubiquity of home improvement stores further attests to a widespread desire to personalize the family home.

In recent decades, some builders have exploited this desire, widening the ability of buyers to modify the product. Under the headline "Express Yourself," one local builder promises the "beginning of a wonderful relationship" via using their "Expression Design Studio," with its staff of interior designers (Shane Homes 2012).

While such choices may still seem superficial, a local "green manufacturer" of movable walls, doors, and floors for offices and homes, who has moved into factory production of houses, attempts to be much more responsive by "producing something in modules that...responds to the design criteria, the functional criteria, the cost criteria and the environmental criteria." In addition to a claimed waste reduction of 99 %, his aim is that "each house...be distinctive...to reflect the family that lives inside it." He uses technology (computer software) to enable clients to "design their own floor plan" and give them "a total experience of what it is they're going to get" (Smed 2012).

10.5.2 Building Design

A major challenge facing architects is that most buildings they design have multiple users – builders, owners, occupants, neighbors, and people in adjacent public spaces – who are all affected by their designs. Although a few clients will pay for a signature building (which will likely have intrinsic value because the signature architect is often successful in their communicating meaning), and a growing number want to minimize energy use (applying criteria like LEED standards), many want nothing more than to maximize their "bottom line." However, the apparent futility of urging clients to fund designing for meaning does not relieve the designer from the moral obligation to do so. One area where designers should vigorously advocate for design which better incorporates meaning, is in the design of public buildings, other facilities, and transportation infrastructure, which sometimes seem to express a deliberate lack of character or meaning.

Historic buildings can also carry a lot of meaning for those users who value tradition, the past, or associate a building with particular historical persons or events. A small group of architects specialize in preservation, conservation, restoration, and adaptive reuse, aiming to retain the meaning of buildings, for present and future generations. This can be a very significant part of a meaningful environment. Although there are notable unsubsidized exceptions, success of conservation efforts generally requires substantial support from all levels of government (in the form of zoning, property, and income tax policies).

10.5.3 Urban Environments

Cities are the center of life for more and more of humanity. Urban planners, urban designers, and architects have a crucial role to play in creating environments (and the processes used to design them) which provide meaning and which express shared public values. Designers should advocate environmentally and socially sustainable environments which not only preserve "natural capital" but also nurture the development of autonomous persons, that is, provide environments which nurture human flourishing by representing meaning to the inhabitants.

Functionalist designers and planners have too often thought of urban environment in terms of traffic generation, mobility, circulation, access and egress, density, land use, and infrastructure. These are necessary. But our urban places are not merely tools for satisfying our basic needs. Rather, our urban environments should be rich receptacles of meaning, value, and tradition, which form part of our framework for autonomous self-determination.

Our buildings and our cities are much more than instruments of need satisfaction; they are imbued with meaning. Our conceptions of our urban environments involve more than brick and mortar, houses and streets. The physical artifacts found in urban spaces have meaning beyond their physical functions. Our descriptions of them often involve thick terms that give them value and meaning. We speak, for example, of "cold and sterile" streets, "congenial" town squares, "formal" gardens, "proud and stately" or "imposing" or "dominating" buildings, and "warm and welcoming" houses. For a person's process of self-determination to be successful, they need a meaningful relationship with their urban environments at all scales.

Our urban environments reflect our conceptions of our collective identities, and our conceptions of what is valuable. Our public artifacts reflect values that are, at least to some degree, shared. Our freeways reflect the value we place on mobility, speed, and convenience; our parks and greenways, the value we place on experiences of nature; and our pathways, the value we place on exercise and fitness. Scruton argues that "Only by transforming the world into the visible and tangible record of things rationally pursued, can a man [sic] find a place for himself there; without that place there will be no self to furnish it" (*ibid*). Such is the importance of *place.*

We believe that an urban environment which nurtures human development, by giving a sense of meaningful relationship, should include (i) a sense of community (including opportunities for social interaction); (ii) access to a range of opportunities for work, education, and recreation; (iii) personal security and safety; (iv) freedom from alienation; and (v) expressions of identity.

The latter two are closely tied to Rawls' (2001, p. 59) fifth *primary good*: "The social bases of self-respect, understood as those aspects of basic institutions normally essential if citizens are to have a lively sense of their worth as persons and to be able to advance their ends with self-confidence."

An urban environment which expresses who we are will have more substance in a unicultural society than in a multicultural society. However, each place does have a shared geography as well as some shared history and heritage, stemming from our liberal democratic traditions. Many immigrants, from other cultures, do come because they are attracted by our societal values and may relate to symbols of this tradition more than we expect. And parts of a city can certainly express other subcultural and ethnic identities (e.g., Chinatown, little Italy) or lifestyles (e.g., hip inner city or middle-class suburb).

10.5.4 Natural Environments

Each person clearly has a right to common natural resources, for example, clean air, water, uncontaminated food, and to a supportive nontoxic environment, in which to have autonomy to be able to pursue their private goals. These rights lead to a concern for environmental sustainability. But our concern here is with the *meaning* that many people find in relation to their conception of a natural environment. Numerous urbanites relate to nature incorporated into the city in the form of trees, greenery, natural parks, greenways, and even streetlighting which allows stars to be visible at night. Others find more meaning in getting "back to nature" by leaving the city.

As a social construction, the concept of "nature" has different meanings for different cultures: aboriginal and nonaboriginal peoples understand and view the natural world in radically different ways (Willems-Braun 1997); third world cultures and first world cultures often have mutually exclusive conceptions of the natural environment (Guha 1999). Even within the Anglo-European tradition, the myth of ancient forest has different associations within German and English mental landscapes (Schama 1995).

The concept of nature also changes over time in the same culture. Before the mid-nineteenth century, the Western Judeo-Christian conceptualizations of wilderness saw it as deserted, savage, dangerous, and desolate in the Anglo-European mind. The Romantic Movement created a great shift in the Western cultural attitude toward wilderness, from the place where one would most likely encounter darkness and despair to the notion of landscape as sacred and sublime: "those rare places on earth where one had more chance…to glimpse the face of God" (Cronon 1999, p. 373). There are also class-based and urban/rural differences in conceptions of nature.

Today, many subcultures (social groups) in North America attach different meanings to widely differing encounters with nature as they conceive of it. These include a vast range of activities, for example, parking an RV in a fully serviced campground, hiking groomed trails in the mountains, climbing mountains, snowshoeing, cross-country or downhill skiing, skidooing, trekking through uncharted "wilderness," bird-watching, nature photography, hunting, fishing, kayaking, sailing, speedboating or sea-dooing on rivers or lakes, and even driving trucks across muddy landscapes. One of the tensions of planning for natural environments is that people who engage in some of these activities consider others of these activities to be a desecration of their conception of nature.

Even more difficult for regional planners is that exploitation of natural resources for economic benefit – petroleum exploration and production, mining, logging, and even agriculture – is often in direct conflict with the types of meaningful "recreational" activities just listed. And these economic activities are frequently incompatible with each other. In making such trade-offs, the fact that nature has significant meaning to these many different users must be taken seriously.

10.5.5 Participation in Planning

Public participation in planning and public design offers great potential for creating more meaningful environments. Two benefits are usually claimed: (i) better planning outcomes and (ii) strengthened democratic institutions. To this we would add (iii) making community or natural environments more meaningful to residents or users by providing opportunities to be involved in planning them. User participation can help overcome some of the alienation we have discussed by giving the occupant/resident a sense of connection to both the designer and the object of design. Participation also enables the planner or designer to get a better sense of what is valued by users and to express it in the design or plan.

Although modernist land-use planning (at least in England) began c.1850 as a political response to the unhealthy conditions of the mechanizing Victorian industrial city, there was always some focus on the meaning of environment to its inhabitants. For example, Howard's Garden City (1902) aimed to bring the countryside (nature) into the city; industrial philanthropists such as Owen and Cadbury built their new towns in the countryside, believing that people would be more human if they regularly encountered nature.

When planning began as a profession (1914 in England), it sought to be more scientific, to develop normative necessary conditions for a "good city." However, planning was by no means completely dominated by scientism, for example, Mumford's (1938) vision was definitely not. He defined community as "people united by common feeling for landscape, literature, language, folkways." He stressed the importance of individual autonomy in that people acting "out of self-respect and respect for other regions, contribute to planning." The value of their contribution came "out of [the] authority of own understanding" (*ibid*), in sharp contrast to the modernist stress on the instrumentalist authority of expertise.

It was only after the success of logistics planning in World War II that the profession became really scientistic, with the "Rational Comprehensive Planning Model" (Harper and Stein 2006, c. 2) dominating theory until the mid-1970s, and practice well into the 1980s, with a continuing influence, particularly outside the "first world."

At least partly inspired by Jane Jacobs' successful opposition to a Robert Moses-planned inner-city New York expressway (1961), planning theorists began advocating some form of public participation in the 1970s. (Advocacy planning, transactive planning, progressive planning, equity planning, and communicative planning all involve some degree of public participation.) Information technology can increase the efficacy of participation by providing better information and the opportunity to visualize different outcomes (Levy 2011).

With various reservations, public engagement in some form has become entrenched in much of North American planning. Although engagement has often been tokenistic or manipulative, its widespread acceptance offers urban and regional planners some scope in making environments more meaningful.

10.5.6 Community

The experience of simply belonging to (feeling part of) a community can really help to overcome individual alienation as well as building social networks (Innes and Booher 2002) and rebuilding "social capital" (Putnam 2000). This can be an additional (fourth) value added to the benefits of participation, when communities of interest and geographic communities are meaningfully involved in planning, design, and other civic affairs. For example, in our home city, 136 community associations (most run by volunteers) are a primary vehicle for citizen engagement in city planning and development processes, supported by a Federation with two full-time professional planners. These associations help to develop a sense of community and of place. At the regional and provincial or state levels, many volunteer interest groups (e.g., fish and game, wilderness preservationists, various environmentalists) participate, although some provinces and states have much better developed formal processes for involving these stakeholders. In some provinces and states, many different forms of public participation have been tried at the regional level, with varying degrees of success (Innes et al. 1994; Innes and Booher 2010; Margerum 2002).

10.6 Conclusions

Perhaps more than we realize, what we experience today still represents the unfolding of the ideas of the eighteenth-century Enlightenment. The humanistic strand asserted the moral standing of the autonomous person and has stressed the importance of authentic identity. The scientific strand led to an enormously improvement in the material well-being of people in advanced (first world) societies. Unfortunately, it also spawned scientism, with its claim that the scientific method is the only source of knowledge and that rationality is limited to the instrumental – the best means to ends (which are seen as nonrational).

The dominance of an instrumental view of people and environments has often resulted in their being treated as objects. As technology has made a wider range of goods available to increasing numbers of consumers, and as they become further separated from the design and production of consumer goods, people have lost their feeling of *connection* to their material environments. This separation has lessened feelings of *meaningful relationship* to their artifacts and their environments, making them feel objectivized as manipulated consumers, that is, less fully human.

With minimal awareness of it, designers and planners have played a significant role in this process of dehumanization. We have argued that there is an ethical responsibility to resist the process and to reassert the value of persons, by designing and planning in ways that increase the meaning of artifacts and environments to users. We have pointed to a few possibilities of doing this at different scales of design.

Notes

1. Some writers, for example, Taylor (1991), consider this second change to be the outworking of modernism.
2. Pragmatism shows that distinctions such as reality/appearance, truth/opinion, objectivity/subjectivity, and fact/value can still be used, when seen as end points of continua, rather than absolute dichotomies
3. Parts of this section (and the next two sections) are based on Badke and Walker (2007). Walker was a colleague in the Faculty of Environmental Design in Calgary for many years and taught design theory with Stein.
4. On the Enlightenment influence, see Friedmann (1987). He views the dominance of "market rationality" as another key aspect.
5. Parts of this section (and the next two sections) are based on Harper and Stein (2006, c.2).
6. Steps of the scientific method: observation of regularities, generalization, theorizing, hypothesis-testing, establishing scientific laws, and uniting theories under general theories. Kuhn (1970) demonstrated that the actual process was more complex than the simple textbook representation.
7. For example, Marxism claims that social structures and moral beliefs are determined solely by economic forces. Reflecting the influence of modernism, the investigation of human and social phenomena eventually came to be known as "social science." In the social sciences, approaches which seek to develop metanarratives are sometimes called "structuralist."
8. Creatures fulfilling these criteria do not necessarily have to be Homo sapiens.
9. A person's initial formulation of a good and meaningful life is largely socially determined. We can concede such communitarian claims, without in any way weakening the moral and political conception of the "autonomous person." But it doesn't follow from any social origin of our goals that we should switch from the individual to the community as the proper object of moral concern.
10. This is the core belief of liberalism, used in a broad sense which encompasses most of the political spectrum in many societies with Anglo-European roots. Liberalism and its notion of the individual have been widely criticized. For an extended explication and defense of liberal ideals, see Rawls (1993; 2001).
11. This same point can be used to demonstrate why a work of art can have intrinsic value.
12. We have not included a discussion of the responsibilities of product/industrial designers because it was beyond the scope of this book. For an excellent treatment of this topic, see Walker (2011) or Badke and Walker (2007).
13. The federal Wartime Housing Corporation built housing units in areas with shortages due to war efforts. Over 45,000 units were constructed from 1941 to 1949. They were noted for the homogeneity of their original appearance (CMHC, n.d.).

References

Badke C, Walker S (2007) Designers anonymous. In: Presented at the education symposium of the ICSID/IDSA connecting '07, World Design Congress in San Francisco, October 2007

Bloom A (1987) The closing of the American Mind. Simon and Schuster, NYC

CMHC (n.d.) Government housing and mortgages in Canada. http://www.canadamortgage.com/articles/learning.cfm?DocID=37. Accessed 25 June 2012

Cronon W (1999) The trouble with wilderness; or, getting back to the wrong nature. In: Desjardins J (ed) Environmental ethics: concepts, policy, theory. Mayfield Publishing Company, Toronto

Dolnick E (2010) The clockwork universe: Isaac Newton, the royal society, and the birth of the modern world. HarperCollins, New York City

Ellul J (1967) The technological society. Vintage Books, New York
Friedmann J (1987) Planning in the public domain: from knowledge to action. Princeton University Press, Princeton
Friedmann J (1993) Toward a non-Euclidean mode of planning. J Am Plann Assoc 59(4):482–485
Galloway MB (1978) User adaptations of wartime housing. Masters degree project, Faculty of Environmental Design, University of Calgary
Guha R (1999) Radical American environmentalism and wilderness preservation: a third world critique. In: Desjardins J (ed) Environmental ethics: concepts, policy, theory. Mayfield Publishing Company, Toronto
Habermas J (1984) The theory of communicative action. Beacon Press, Boston
Harper TL, Stein SM (2006, reprinted 2012) Dialogical planning in a fragmented society: critically liberal, pragmatic and incremental. Transaction Publishers, New Brunswick
Howard E (1902, republ 2009) Garden cities of to-morrow. Routledge (orig. publ. 1898 as To-Morrow: a peaceful path to real reform)
Hummel RP (2008) The bureaucratic experience: the post-modern challenge, 5th edn. M.E. Sharpe, London
Innes J, Booher D (2002) Network power in collaborative planning. J Plann Edu Res 21(3):221–236
Innes J, Booher D (2010) Planning with complexity. Routledge, London
Innes J, Gruber J, Neuman M, Thompson R (1994) Coordinating growth and environmental management through consensus building. California Policy Seminar, Berkeley
Jacobs J (1961) The death and life of great American cities. Random House, New York City
Kant I (1785, republ 1958) Groundwork of the metaphysic of morals. In: The moral law (trans: Paton HJ). Hutchison, London
Kuhn TS (1970) The structure of scientific revolutions, 2nd edn. University of Chicago Press, Chicago
Lansley S (1994) After the gold rush: the trouble with affluence. Century, London
Le Corbusier (1923) Vers une architecture. Recent English version (trans: Goodman J) (2007) Towards an architecture. Getty Publications, Los Angeles
Levy RM (2011) Virtual reality: a tool for urban planning and public engagement. Computers in urban planning and urban management (CUPUM). In: Proceedings, 7 July 2011, Lake Louise, Alberta
Margerum RD (2002) Evaluating collaborative planning: implications from an empirical analysis of growth management. J Am Plann Assoc 68(2):179–193
Mumford L (1938) The culture of cities. Harcourt and Brace, New York
Murdoch I (1970) The sovereignty of good. Routledge and Kegan Paul, London
Nozick R (1974) Anarchy, state and utopia. Basic Books, New York
Oskamp S (2001) Rampant consumerism: a public and environmental health threat. Monit Psychol 32(4):52
Putnam RD (2000) Bowling alone: the collapse and revival of American community. Simon and Schuster, New York
Rawls J (1993) Political liberalism. Columbia University Press, New York
Rawls J (2001) Justice as fairness: a restatement. Harvard University Press, Cambridge, MA
Rorty R (1981) Philosophy and the mirror of nature. Princeton University Press, Princeton
Rorty R (2000) Universality and truth. In: Brandon RA (ed) Rorty and his critics. Blackwell Publishers, London
Rowland W (1999) Ockham's razor: a search for wonder in an age of doubt. Key Porter, Toronto
Rühe P (2001) Gandhi. Phaidon Press Ltd, London
Schama S (1995) Landscape and memory. Alfred A. Knopf, New York
Scruton R (1979) The aesthetics of architecture. Princeton University Press, Princeton
Scruton R (2011) Green philosophy: how to think seriously about the planet. Atlantic Books, London
Shane Homes (2012) Express yourself. New Homes. Calgary Sun, special edition, 7
Smed M (2012) What I know about modular construction. Avenue Magazine. Redpoint Media, Calgary, 42

Sparke P (2006) EVDS keynote series lecture, given at Calgary chamber of commerce, 22 February 2006
Stein SM, McMordie M, MacKinnon L (1999) What's wrong with … 'What's wrong with plastic trees'?
Taylor C (1991) The malaise of modernity. Anansi Press, Concord
Walker S (2002) Objects as symbols of beauty. http://www.idsa.org/sites/default/files/2002_Stuart%20 Walker.pdf. Accessed 15 May 2012
Walker S (2011) The spirit of design: objects, environment and meaning. Routledge/Taylor and Francis, London
Willems-Braun B (1997) Colonial vestiges: representing forest landscapes on Canada's west coast. In: Barnes TJ, Hayter R (eds) Troubles in the rainforest: British Columbia's forest economy in transition. Western Geographical Press, Victoria
Williams B (1985) Ethics and the limits of philosophy. Fontana Press, London

Chapter 11
Risk, Space, and Distributive Justice

Claudia Basta

11.1 Introduction[1]

In this introduction, I will start my contribution by recalling a personal experience. Disconnecting my ideas from that experience would critically impoverish my capacity to convey my ideas' underlying motive and, I believe, their significance.

The episode dates back to 2009 and to a conversation I had with some members of the then a Dutch advisory committee on hazardous substances. I had invited them to provide feedback on a project I was about to submit to a scientific grant competition. The project intended to combine selected risk analysis methodologies, spatial planning theories, and ethical theories, so as to elaborate an integrated evaluative framework for the siting of impacting and hazardous technologies. My starting point during this discussion was the lack of an integrated perspective on the matter of siting such technologies – a perspective that could support decision makers, and planners with them, in addressing "technical" and "nontechnical" considerations consistently throughout the entire evaluative process. Typically, techno-economic appraisal that guides the selection of different candidate sites is frequently "insulated" from relevant ethical implications. That seemed the firmest point to show why resistant appraisal models were destined to failure, at least in some regards. I then argued that the applied research in the field should have advanced more value-sensitive appraisal models, stressing that the resulting "value-sensitive" appraisal model would have not entailed losing any evaluative rigor. In essence, what I proposed was solving the disjunction between the analytical *and* the normative components of siting evaluations by applying one and the same ethical theory throughout the entire evaluative process. From risk appraisal, sites inventory, site selection, and land use planning in

---

C. Basta (✉)
Land Use Planning Group, Environmental Sciences Department,
Wageningen University and Research Centre, Wageningen, The Netherlands
e-mail: Claudia.basta@wur.nl

C. Basta and S. Moroni (eds.), *Ethics, Design and Planning of the Built Environment*,
Urban and Landscape Perspectives 12, DOI 10.1007/978-94-007-5246-7_11,

the area surrounding the chosen site, the theory I proposed was the theory of distributive justice of John Rawls (1971). This, I explained, was a main theoretical current within the Ethics of Technology group of Delft University of Technology where I had started to develop my postdoc project; therefore, I closed my presentation by emphasizing how my intent, as spatial planner, was connecting this prominent philosophical current with the domain of technological risks, and specifically with the matter of siting risky installations.

The experts listened attentively to my enthusiastic presentation and provided valuable input and, at times, helpful critical remarks. Then, I was finally encouraged to proceed with my application. But when I was about to leave the meeting, one of the experts dismissed me by saying "Indeed your research on justice is truly needed. We need to improve our risk communication and *letting people understand* that we do perform the most consistent risk analysis and site selection each time we need to take a sensitive decision." Or something to that effect.

In that precise moment, I understood what had happened during my presentation. In the best scenario, I had been perceived as the one more idealistic scientist with intricate, yet kind of fashionable ideas; in the worst scenario, my proposal had been filtered by discarding all theoretical intricacies at the end of identifying its possible "practical," if not instrumental, use. I think what happened was the latter. The term *consistent*, for example, was disconnected from my intended meaning and perceived as a smart attribute to associate to risk analysis; *sensitive*, by its side, was perceived as a very effective term to be conveyed to the public in relation to "decisions." That the evaluative intent of my framework was to provide a form of *ethical* consistency that would affect *the course* of those decisions had been totally overlooked.

While realizing this, I surrendered to the impossibility of clarifying my ideas any further – these ideas were, evidently, premised on a worldview entirely different from my audience's. Moreover, it was a worldview which had no chance of being comprehended in its unity without an instinctive distinction of grouping it under the rubrics of "this is of practical use" vs. "this is just academic speculation." And indeed I left the meeting wondering, for a moment, whether I was only speculating along a fruitless direction.

This experience revealed a fundamental thought. Both my project and the authoritative audience of that meeting were animated by the same, sincere, intent: refining the evaluative instruments at disposal in the domain of hazardous facilities siting at the benefit of society. However, the words of the expert who accompanied me out of the meeting forced me to realize that the ethical trajectory signed by my project had no chances of penetrating tested governmental evaluative models beyond some superficial infiltrations. Differently put, this episode forced me to realize that the approach to the siting of hazardous installations of governmental experts, in such an enlightening way summarized by the remark "we need to let people understand," in the framework of my research should have been part of the very problem.

This predominant governmental approach could be illustrated in many ways, and perhaps the most effective way is what Owens (2004b) calls the "techno-rational" model of appraisal. Here, the decision maker acts as promoter of the public interest by grounding her decisions on the solid terrain of technical assessments. Such assessments are usually provided by supposedly "neutral" scientific advisors. The third party, that is, the individuals affected by the decisions, are involved in the decisional process only later on, at the end of being persuaded of its "rightness" through extensive explanations of its objective rationale. As the experts put it, "we need to let people understand"; that is, we need to persuade, to convince; we need to enlighten the darkness of subjectivity with the brightness of rationality. Subjectivity, in short, is a problem to solve – not the epistemological source through which giving form to the problem itself. So how could I have pushed my ideas through? Their essential was that without a shared ethical point of reference that could encompass different worldviews, the very decisional problem could not be formulated. The remarks of the expert, despite motivated by the same good end, summarized with insightful simplicity that one worldview had to prevail on the other instead.

I am aware of the comparative banality of remarks like these. Literature has long since identified the irresolvable epistemological opposition that underlies the dialogue between parties called to interact on controversial siting decisions (Huitema 2002; Boholm 2004; Boholm and Lofsted 2004; Owens 2004a, b; Hayden Lesbirel and Shaw 2005). Several authoritative arguments have been already advanced in the attempt to solve the polarization between, particularly, risk posers and risk runners by orienting the discourse on risk toward the identification of the moral rights and moral obligations of each (Hansson and Peterson 2001; Peterson 2003; Peterson and Hansson 2004). Others have extensively argued on the fallacious distinction between a "neutral" and a "biased" perception of risk, advancing the metaethical argument that the notion of "risk acceptability" cannot be deprived of its moral emotional component (Roeser 2006 and Roeser 2010 in particular). Planners, in this latter respect, provided valuable empirical inputs to the discussion by documenting matters of self- and place perception among citizens living nearby hazardous installations (Simmons and Walker 2004). As I will elaborate in the following sections, these latter inputs enriched the theoretical debate on "risky siting" by highlighting the need to always consider the site-specific implications of concrete installations, so as to prevent overarching generalizations incapable to reconnect the theory of siting to its practice.

Literature, in short, abounds with contributions of a growingly multidisciplinary scholarly community, to the point that adding to it seems less productive than taking a stance within its various currents. However, it is precisely in light of this abundance that discussing why the techno-rational model of appraisal keeps guiding the siting of impacting and hazardous technologies becomes a matter of importance. In this contribution, I will address some considerations precisely in this direction. To do so, I will refer to a recent case of rejection of a CO_2 underground disposal by the side of the citizens of Barendrecht (the Netherlands). This case was discussed, under a different light, in a previous contribution (Basta 2011). Here, I will concentrate on what I do regards as the aspect of the governmental decisional approach that led the course of events to conclude with fierce societal opposition. I propose that this aspect relates to the lack of distributive considerations of the impacts and risks of the installation during the phase of appraisal of the "suitable" site among the considered candidate sites. The next section clarifies the background of this position by recalling some prominent theories, namely, the sociological discourse on risk of Ulrich Beck (1992) and the theory of distributive justice of John Rawls (1971), the latter transposed to planning theory by Stefano Moroni (1994, 1997).

11.2 Technological Risks and Distributive Justice: Some Theoretical Premises

Since their appearance in the English translation of 1996, the ideas of the sociologist Ulrich Beck on the advent of a risk society kept animating, at times critically and at times unreservedly, the debate on technological risks (Beck 1992). As is well known, Beck's seminal writing revolves around the observation that the Western economic development pattern, in late modernity, was primed by the massive introduction of hazardous technologies of yet unknown risky potential. Such technologies create forms of inequality in society not only regarding individuals' access to the produced benefits but also in relation to their exposure to the relevant risks (Beck 1992). Throughout his first academic best seller, Beck does not make distinctions between different types of risk, referring to a general notion of it while mentioning a number of (technological) examples. This is an important remark in the context of this contribution: as noted, "pollutants in foodstuff" and a nuclear factory are risk sources whose specifics and consequences are remarkably different, if not incomparable (Leiss 2001). However, if we could rewrite parts of Beck's first book by adding "large-scale and site-specific" before many of the "risks" he mentions, this room for critical remark would vanish. The only necessary distinction at that point would regard the technological risks "not tight to the place of origin" (Beck 1992, p. 22) because of their intergenerational relevance and the risks "tied to it" because of the intra-generational implications of their consequences. With a good deal of simplification, we could say that technologies like nuclear installations and CO_2 underground disposals belong to the first category, whereas chemical and energy installations belong to it depending on the specifics of the involved substances, operational standards, and the effects of the relevant risks on man and the environment.

Recalling Beck's work is therefore only meant to highlight the following: site-specific hazardous technologies create situations of unequal risk distribution in society. Approaching the matter of their siting requires, consequently, to address considerations of fairness and equity (Keller and Sarin 1995; Davy 1996; Linnerooth-Bayer and Löfstedt 1996; Hayden Lesbirel and Shaw 2005; among others).

To elucidate these two notions, I will refer to John Rawls' authoritative *Theory of Justice* (1971). As is well known, the pillars of the Rawlsian theory are the conception of "primary good" and the two principles of justice. The former refers to the goods each individual would be entitled to, in accordance with a contract that would regulate individuals' coexistence within a fair society. These goods would be identified by individuals acting behind the "veil of ignorance," that is, without knowing what their social, ethnic, and religious statuses would be in a real setting.

Arguably, in this original position, individuals would identify as "primary" the system of liberties and essential means they would need in order to develop the own self freely. Such liberties and means are therefore those that can be equally assigned to anybody without violating others' liberties. Freedom of speech of religious orientation and access to basic education are some examples. The two principles of justice relate to this conception of primary goods by, respectively, stating the obligation of their equal distribution in society (1st principle) and by legitimizing inequalities only when they do benefit the whole, up to its most disadvantaged members (2nd principle). The distinctive aspect of Rawls' conception of justice therefore aims to prime distributive mechanisms of primary goods that would put all in the same initial condition, instead of aiming at a restrictive form of social equality. In more simple words, the fair society the philosopher had in mind is not a society in which everybody is equal in terms of status and belongings, but rather a society in which everybody has access to the essential rights and means needed to develop their own aspirations. Rather than a radical conception of equality as "leveling differences," the liberal conception of justice put forward by Rawls is therefore of equality as "departing from the same set of essential means." That is why Rawlsian theorists identify justice *in* fairness.

Beck's risk society has interesting elements of affinity with this theory. If site-specific technological risks create a threatening form of inequality in society, it follows that it can be envisioned either a moral obligation of removing such risks or of justifying their uneven distribution only when it benefits the whole. This "moment of connection" between the two theories is surely interesting, as among other implications it gives form to the problem of *planning* a fair risk distribution in society.

The friction between the analytical and the regulatory dimensions of "risky politics" and the challenge of reasserting justice in a risk society has been touched by law, sociology, and political philosophy writings (Arcuri 2005; Huitema 2002; Hudson 2003). Individually, the two theories have substantially penetrated the planning literature. The Beckian discourse on risk infiltrated it through conceptions other than specifically technological risks, particularly the notion of "reflexive modernity" and the consequent horizons of a reflexive planning discourse (Howe and Langdon 2002). Other planners concentrated on the "ideology of certainty" that keeps permeating planning systems despite "risky realities" started to be acknowledged as the only possible realities (particularly, Gunder 2008). Rawlsian lines of inquiry, by their side, penetrated the planning literature in the works of Harper and Stein (1992), Moroni (1997), Stein and Harper (2005), and Fainstein (2010). In light of the scope of this contribution, I will elaborate on these Rawlsian lines of investigation in planning theory more at length.

The planning theorist Moroni (1997) argued in details on the implications of the Rawlsian conception of "primary good" for the spatial planning theory. Replicating the line of reasoning of the individuals acting behind the veil of ignorance, Moroni argued that decent housing and the access to green areas and to sufficient transport are the "primary" among the spatial goods. Such spatial goods constitute the basic "spatial" condition each individual should depart from in order to pursue the own objectives of self-realization and happiness. As such, they constitute the basic "spatial rights" that should be guaranteed to each member of society up to the most disadvantaged.

But the main point of interest of Moroni's elaborations is his addition to the original Rawlsian list of primary goods of a "safe living environment." When formulating the negative right of *not* being unsafe, it indeed follows that all members of society are entitled to live in "spatially safe" conditions. Although Moroni does not discuss this point at length, I conclude that this condition includes living protected from major hazards and being exposed to tolerable levels of risk.

This is an interesting point as it relates to the matter of siting hazardous and risky technologies and planning the surrounding urban areas in accordance with both the Beckian vision of unequal risk distribution in society *and* the Rawlsian conception of justice as fairness. From a planning perspective, the resulting disciplinary implication is that aim of the planning practice should become distributing spatial safety in society equally up to its most "spatially disadvantaged" members. The main evaluative implication, consequently, consists of identifying the level of spatial safety individuals are universally entitled to. Arguably this level has a concrete geographical dimension (e.g., "distance from" or "emergency routes toward"), and its means of distribution are planning instruments. Because of their rights' and permits' allocation purpose, arguably land use plans are the privileged instruments for realizing such distribution.

These conclusions constitute the lenses through which I will discuss the case of the CO_2 underground disposal proposed for siting in Barendrecht. The essential viewpoint enabled by these lenses is the following. Asserting that spatial safety is a primary good, and that there is a level of it that should be equally guaranteed in society, equals stating that safety *and* fairness are, respectively, the *dimension* and the *criterion* that should guide planning decisions in relation to risks and hazards. When thinking at the limitations of the techno-rational model of appraisal described above (Owens 2004b), this position provides a promising evaluative perspective. Through it, siting decisions are approached as decisions that *ought to* embed, next to a sound assessment of the site-specific impacts and risks of hazardous installations, distributive justice considerations. This point will become particularly pertinent in the following section.

11.3 The Failure of the Techno-Rational Notion of "Suitable Site": A Recent Case

The case of the CO_2 underground disposal proposed for siting in the Dutch town of Barendrecht was discussed, under a different light, in a previous contribution (Basta 2011). Here, I will discuss it in relation to the current theoretical premises, particularly the principle of fair distribution of technological risks through spatial planning instruments.

In the course of 2009, the municipality of Barendrecht, in the southern part of the Netherlands, engaged in an open conflict with the Dutch government in relation to the proposed siting of a CO_2 underground disposal. The government promoted the project in concert with a known (inter)national oil corporation, which was requested to implement carbon capture and storage technology in the framework of the national policy on climate. This technology allows to capture carbon emissions at source and storing them underground into exploited gas fields, of which the Netherlands are particularly rich.

In order to assess the best possible location for the disposal, the Dutch government delegated a sites inventory study to the national Applied Research Institute (in the following, TNO). Twelve possible locations, of which seven were offshore and five inland, were considered (Breunese and Remmelts 2009). Among the 12 candidate sites, the municipality of Barendrecht was the more densely populated. Nevertheless, the geological characteristics of the gas field underneath its surface and some key techno-economic constraints (principally, the length of the pipelines connecting the points of emission of nearby refineries to the storage) led the research agency to conclude that Barendrecht was a suitable site for proceeding with the pilot project. The government defended this outcome of the inventory study by recalling the national interest, alleging that "capture and storage of CO_2 is a necessary transition technology to help cut carbon emissions" (Reuters, November 18, 2009).

The local population opposed the pilot project, which should have led to the installation of the disposal within the following few years, on the ground of its feared impacts on property values and, more generally, on the image of the town (Terwel et al. 2012). The slogan "we do not want to become the national CO_2 dumping place" spread through manifests, interviews, and public consultations. Following the growing opposition at a local scale, the debate on the technology option of CO_2 underground disposals became of national resonance. As a result, several Dutch municipalities declared to reject the possibility of being future candidate sites for similar disposals. Somehow, an initially local case of opposition to a specific installation became a sort of national movement against the very carbon capture and storage technology. However, the Dutch government kept defending the project, justifying its desirability in light of the national policy objectives of climate change response (Dutch Ministry of Environment VROM 2007).

I think it is evident that the story line of the Barendrecht case recalls the typical "techno-rational" model of appraisal discussed above. The "owner" of the decision-making process was the national government who, consistently with a specific policy objective, delegated the inventory of possible sites to a supposedly neutral advisory body. The latter assessed a number of candidate sites on the basis of several criteria, which included the geological characteristics of available gas fields, the costs of each alternative, and the relevant technological requirements (Breunese and Remmelts 2009). Among the candidate sites, the area of Barendrecht was the most densely inhabited; nevertheless, the main reason for its indication as suitable site seemed to be the criterion of cost-effectiveness, which is explicitly related to the distance between CO_2 sink and sources. The distance between the exploited gas field of Barendrecht and the source of CO_2 emissions was estimated about 20 km; all other sites were in the range of 75 up to 210 km. Here, the report reads "the fact that the Barendrecht field is located under a built area is not of a (geo)technical nature and therefore impossible to weigh against the other geotechnical factors considered"

(Breunese and Remmelts 2009, p. 26). That is to say, the criteria considered by the research agency at the end of formulating a judgment of suitability were, allegedly, only geotechnical and cost-effectiveness criteria. More explicitly put, the characteristics of the Barendrecht' gas field and the 20 km of pipelines connecting the nearby area of Pernis to it were given the highest weight in the site inventory study. The immediate conversion of the relevant outcomes in the final siting decision makes even reasonable to consider whether the inventory had been merely instrumental to provide "a post-demonstration of a preconceived judgment" (Owens 2004a, p. 1946). Somehow, Barendrecht seemed predestined to host the disposal. This hypothesis, however, would leave the main point of our discussion unchanged, as this "post-demonstration" would have also been constructed only on the ground of a techno-rational epistemology.

Evidently, this epistemology excluded the consideration of the instances of a densely populated locality chosen among other non-inhabited candidate sites. Recent accounts confirmed that the main concerns of the population of Barendrecht regarded the feared impacts of the installation on the image of the locality, the consequent fall of property values, and the possible risks brought by the technological installation (Terwel et al. 2012). It is important to keep in mind that due to its vicinity to the major industrial area of Pernis (i.e., to the source of CO_2 emissions of the proposed storage), Barendrecht is already "featured" by the massive presence of industries and refineries. Under this light, the slogan of citizens "we do not want to become the national CO_2 dumping place" seems to reveal the underlying motive of rejection, that is, the opposition to a worsened condition of impacts and risk exposure but also to a further negative stigmatization. Public acceptance was therefore early recognized as the key challenge of the project also from the side of the international oil corporation (Kuijper 2011). The late involvement of citizens in the decisional process and a subtle mistrust in the underlying motive and reliability of authorities played also a significant role in shaping the conflict (Brunsting et al. 2011).

This could be obviously reduced to a "Not In My Back Yard" (NIMBY) story line: that is, *not here – not now*. However, under a different perspective, we could consider the already high concentration of industries in the area as a key factor of opposition. Barendrecht saw itself chosen among others, uninhabited candidate sites on the ground of purely techno-economic considerations. Should other types of considerations, for example, considerations of a fairer impacts and risks distribution at regional or national scale, having been considered instead?

I think that this case exemplifies the resistant techno-rational model of appraisal and its lack of permeability to such considerations in a paradigmatic way. The crown on this paradigm are the words of the proponents of the CO_2 installation, pointing out the need to design "a comprehensive public acceptance strategy" (Kuijper 2011). That is, "public acceptance" should have followed the outcome of the sites inventory as an object of strategic design; the underlying rationale, in essence, is that public acceptance is something to obtain ex-post through strategic maneuvers under the "we need to let people understand" motto.

Questions of moral relevance, and of competitive ethical perspectives, do therefore inevitably arise. I will discuss them in the following section.

11.4 Discussion: The Suitable and the Fair Site – Irreconcilable Realities?

The main question that stem from the case described above is, first of all, whether it is morally justifiable to create conditions of risk inequality in society despite the theoretical availability of alternatives. Furthermore, the case leads to question whether the risk inequality created by site-specific technologies whose benefits are of global relevance is always justifiable in light of the second Rawlsian principle of justice; as explained, this principle states that societal inequalities in primary goods distribution can be justified only when such inequalities benefits the whole up to the most disadvantaged.

Without digging into the technicalities of each alternative and into the specifics of the case under discussion, I will provide tentative answers while indicating some main open questions. First of all, commonsense suggests that if among the criteria used to review the 12 sites of this case the criterion of "not worsening preexistent conditions of impacts and risk exposure" would have been considered, the site of Barendrecht could have been discarded during the site inventory or even not being considered at all. On a marginal note, this does also allow to suggest that the following national debate on carbon capture and storage as a technology option for climate change response could have developed toward a different direction, possibly without polarizing Dutch citizens and the Dutch government into two irreconcilable positions.

But it is the "not considered at all" point that is of paramount importance here. That is what exemplifies how addressing distributive considerations at the early stage of the planning of technological risks could affect the entire course of siting decisions – and providing, at the same time, a morally solid framework to all possible successive confrontations with public resistance. Evidently, this moral solidity rests on a shared notion of justice among the involved individuals.

To discuss the case above using the lenses of my theoretical premises, it is fundamental to capture that what generates such confrontations are, indeed, different notions of justice. What "fairness" entails is often object of disagreement among individuals, to the point that "competing views about fairness are at the core of the siting impasse" (Linnerooth-Bayer and Fitzgerald 1996). This is brilliantly argued by Davy (1996) in his account of the "justices" that compete during siting processes. Davy unveils the essentials of the main three justices among them, namely, the utilitarian justice that informs the techno-rational model of appraisal (providing for the least dissatisfaction or the least risk), the libertarian notion of justice (minimizing state intervention and enhancing competitive interactions), and, finally, the Rawlsian or egalitarian notion of justice (that, as discussed, allows inequality only if it is beneficial to the whole, including the most disadvantaged). All three conceptions of justice are justified by either some underlying epistemological position (like the techno-rational epistemology of utilitarianism) or by some underlying position relating to normativity (like the sphere of intervention of the state vs. the sphere of liberties of individuals in libertarianism). As these three conceptions of justice coexist during siting processes, "any siting outcome will be unjust or unfair to somebody" (Linnerooth-Bayer and Löfstedt 1996). Davy elaborates on this crucial point by stating that "the problem of the coexistence of different concepts of justice cannot be solved by identifying and pursuing the 'right' concept of justice" (Davy 1996). Taking the "suitable site" to correspond to the "just decision" does rather require an effort of anticipation of all possible forms of perceived and concrete unfairness. In this view, the decisional process should be bent toward minimizing such unavoidable unfairness to the extent possible. In Davy's vision, in essence, competing notions of justice do not call the respective individuals to opt for the one that best suits the specific circumstances; they rather call to a resilient response, by the side of the decisional process, to the inevitable feelings and conditions of unfairness that will accompany it. This is what Davy (1996) calls "justice as compassion".

This is the point of Davy's insightful analysis from which I wish to take a cautious distance. The equal justifiability of different perspectives on fairness does not entail, per se, the obligation to not disabling any of them in the course of sensitive decisional processes, like the siting of hazardous installations. If any notion of justice, regardless of the underlying epistemological or normative position, is equally admitted in the arena of discussion, then all following "injustices" becomes insurmountable obstacles. This remark, observe, does not intend to question the *legitimacy* of different notions of justice; it only intends to question whether *all* of these notions can be equally and simultaneously considered at the end of opting for the just decision.

To clarify this point, I propose to make a distinction between the *fundamental* injustice that derives from the clash of different notions of it from the *residual* injustice that derives from the negotiated option for one of them. The former form of injustice derives from epistemological or normative inconsistencies among the perspectives of individuals; the latter form of injustice instead derives from the sacrifice suffered by some *because of* and *despite* the adoption of a common line of principle.

This can be clarified by going back to the example of the CO_2 underground disposal. Let us assume (for the sake of argument) the proven tolerability of the impacts and risks associated to the disposal and the proven global relevance of its benefits in terms of climate change response. The underlying assumption, to be clearer, is that there exist a rough metric that allows to counterbalance the local impacts of the technology with the global benefits provided by the reduced emissions of CO_2. The global relevance of such benefits would obviously include the citizens of Barendrecht. Let us now suppose that all available siting alternatives are equally inhabited areas, and that the respective urban districts are from equally to more densely industrialized. Following distributive considerations aimed at identifying the most even distribution of impacts and risks among the candidate sites, it could be considered *fair* to site the disposal underneath the town of Barendrecht. Here, the most spatially disadvantaged citizens would equally be among the direct beneficiaries of the installations; therefore, should the location of Barendrecht the one guaranteeing an even distribution of risks among the candidate sites, an egalitarian planning process could opt for siting the CO_2 disposal there. Thereby, land use plans would be regulated accordingly.

This "fair" decision, however, would not prevent neither an increased condition of exposure for the citizens of Barendrecht nor feelings of injustice and resentments among them because of their spatially disadvantaged condition. A condition of *residual* injustice would likely be experienced. However, the principle of fairness that guided the siting decision would provide a different and, observe, *moral* point of reference through which giving significance to such condition. What citizens "would have to understand" would be being equal beneficiaries of the global benefits of the installation *despite* their spatially disadvantaged condition, which would be created because it would not disadvantage further the citizens of other candidate sites. "Understanding" this moral justification of the siting decision would be sharply different than "understanding" the arguments of the techno-rational justice, which would have privileged the choice of the site on the ground of the costs/benefits rationale. The same would apply to the "understanding" of the rationale of libertarian justice, which would have privileged spontaneous sites candidatures and direct negotiation of compensative benefits between risk posers and risk runners with the likely result of siting the CO_2 disposal in the most economically disadvantaged site, thereby worsening a preexistent condition of inequality.

To conclude, while prominent authors agree in identifying the siting impasse in the coexistence of different notions of fairness, there is room for disagreement in Davy's indication of not pursuing the "right" notion of it prior to guide siting processes to a possible just conclusion. Without an early agreement on what fair is, and what this fairness entails for the involved parties in terms of (possible) fundamental injustices and (inevitable) residual injustices, there is no room for a morally solid decisional process. The lack of such agreement would equal not setting down the rules of the game in advance and then let the actors play freely, thus without a common line of principle through which judging each other's positions, actions, and ultimately residual condition and feelings of unfairness.

This contribution suggests that the fairness pursued by the distributive justice paradigm is the line of principle that can accommodate such residual forms of unfairness more, and more consistently, than any other. From a moral perspective, being the subject of residual injustice in the framework of a decisional processes whose explicit and common principle of fairness is the equal distribution of impacts and risks of technological installations is totally different than being subjected to injustice because of the exclusion, a priori, of explicit and early agreed upon principles of references. In support of this argument, I suggest that the egalitarian position is the real device of the sentiment of "fairness as compassion" that Davy refers to in his work (Davy 1996). With it, Davy intends the compassionate consideration of all possible notions of justice, and implicitly of the sense of injustice suffered by some, as worth of equal consideration: "[fairness as compassion] advises to consider each of the different notions of justice that are involved in LULU and NIMBY disputes and to eliminate and avoid injustice to the extent possible" (Davy 1996, p. 107). This is a valid indication. However, a possible different connotation of "compassion" in the domain of hazardous facilities siting could be that of "accepting residual injustices" as part of what individuals owe to each other within the human community. In increasingly urbanized realities, characterized by growing complexity and increasingly intrusive infrastructures, to the one who will have to cope with the proximity of a CO_2 storage, the one who will have to cope with the vicinity of high-speed rail connections, renewable energy installations, or major power lines corresponds. Fairness as compassion, rather than an inclusive attitude of proponents of installations and decision makers toward different perspectives on fairness, is primarily the sentiment that should animate the individuals subjected to uncomfortable siting decisions. Such sentiment can only flourish from a shared guiding principle of fairness and from the reasoned acceptance of the individual share of sacrifice this guiding principle will residually, and inevitably, entail.

Arguably, rooting this sentiment in reality is the real challenge of any hazardous facilities siting processes. However, it is less likely to see this sentiment flourishing when such processes depart from a fundamental disagreement on what makes a siting decision a fair decision. Probably, the counterpart of this sentiment, that is hostility, is destined to become the protagonist all the times that through early, inclusive, and active societal participation such agreement is not even attempted.

11.5 Conclusions: Some Critical Notes for Future "Fair Risk" Planners

In this chapter, I tried to outline some undergoing reflections on the matter of siting impacting and risky technologies by taking different theoretical perspectives together. These reflections were triggered by a personal experience that, while confirming commonly held opinions of prominent interdisciplinary scholars in the field, cemented my determination to continue exploring the horizons for an ethical discourse within spatial planning in at-risk areas. In this contribution, the accent was posed on the ethical perspective offered by the theory of distributive justice of Rawls (1971). Together with Beck's vision of the risk society (Beck 1992), I tried to elucidate the relevance of this theory to the siting of hazardous and risky facilities. In support of my arguments, I referred to the story line of a controversial siting case in a Dutch town. Here, the pilot project of a CO_2 underground disposal was opposed by the local community, already living nearby a heavily industrialized area, following concerns regarding its impacts on the locality and fears of further stigmatization. The aspect of this case on which I draw more attention is precisely this preexistent condition of exposure of citizens to major risks and hazards the siting of the CO_2 installation would have inevitably worsened. I argued that this preexistent condition should have been an explicit criterion of the sites inventory carried out by the Dutch government. I therefore suggested that distributive considerations regarding the impacts and risks of site-specific technologies should inform "fair" siting processes and, particularly, the selection of the "suitable" site among different candidate sites.

Not many of the ideas I proposed here were original ideas. Their most original aspect is my attempt to develop them within the planning discourse as much as into the ethical discourse. The implications of my discussion for the planning theory and practice will therefore be the focus of my concluding remarks.

One of the premises of this contribution was that in the domain of hazardous facilities siting land use plans are distributive instruments meant, on the one side, to prevent the violation of the spatial right to safety and, on the other side, to distribute the "spatial burdens" of impacting and risky technologies by striving for conditions of equality. The second Rawlsian principle of justice allows for inequality in primary goods distribution when such inequality benefits the whole up to the most disadvantaged; a point to be clarified is therefore how this principle should apply to the additional *spatial* primary goods that, following Moroni (1997), I proposed should lengthen the original Rawlsian list.

Notably, it was Rawls to leave the list of primary goods open to additions and, furthermore, to suggest that additional primary goods could be object of post-constitutional agreements[2]. Spatial primary goods would intuitively follow in this category of "additional agreements," as arguably their features would vary according to specific geographical and cultural settings.

Next to basic "material" goods such as decent housing and access to green areas, the spatial good of safety is "immaterial" and as such, somehow, more fundamental. The "right to safety", here intended as a precondition of exertion of other fundamental liberties, encompasses all material elements of the built environment. Not being exposed to intolerable hazards, having access to emergency routes and recovery in case of natural or man-made disasters but also accessing buildings and transport routes constructed according to precise and inviolable accessibility and safety standards are examples of the multifaceted implications of the notion of "spatial safety" in the context of our discourse. There is, obviously, no standard of safety that could be equally "distributed" through spatial planning instruments without relating this notion to specific elements of the built environment; there is, rather, an equal "bottom line" of safety that ought to be guaranteed to individuals in relation to each specific and potentially "unsafe" spatial condition. For example, houses are to be constructed according to given and not violable safety standards, transport routes are to be designed according to given criteria of accessibility and emergency response, and so on. In the case of hazardous facilities, this "bottom line" of spatial safety should consist of not exposing individuals to intolerable risks because of an unsafe proximity of hazardous and risky installations. Note that this is long since prescribed by the European common regulation through the Council Directive on Hazardous Substances (96/82/EC and following amendments), and that there is an extensive literature on the relevant experiences of implementation of member states (see Basta 2009 for an extensive account).

From an evaluative but also practical perspective, this objective constitutes a problem of often irreducible complexity. This complexity is due to the fundamental societal disagreement on the desirability and justifiability of certain technological risks, on the one side, and to the situations of proximity between industrial sites and residential districts inherited from our rather "risk deregulated" past on the other side.

Regarding the first point, that is, the evaluative problem of deciding whether a risk is tolerable or not and whether the respective technology is irreplaceable and thus necessary to achieve greater benefits for the human community, the room for disagreement is virtually infinite. It is as infinite as the nuances coloring the notion of justice hold by different individuals. In any concrete setting, however, this is the most important preliminary question to pose. Regarding the specific case discussed here, that is, the technology of carbon capture and storage into underground disposals, it is important to point out that the technology is globally acknowledged to support the transition toward a nonfossil fuel era to the point that the Dutch government considers its implementation as the third pillar of the own Climate Policy (VROM 2007). I do not wish to take a position in regard of neither the "irreplaceability" nor the "necessity" of this specific technology, as this would require to investigate into a different direction and to deviate from the scope of this contribution. However, I suggest to leave the possibility of its transitory large-scale implementation open to rigorous ethical investigation, as I suggest the same openness in regard of renewable energy and water technologies meant to respond to the increasing scarcity of finite resources and to a growing global environmental degradation.

The second reason of complexity regards situations of proximity between impacting and risky industrial sites and residential districts European cities have inherited from the past, and that could further derive from specific technological constraints or scarcity of available sites. Here, I conclude by advocating that a distributive perspective on the relevant evaluations could offer a valid approach to prevent further societal inequalities while providing "a solid moral basis for contemporary planning theory" (Stein and Harper 2005, p. 147). This conclusion is partially in contrast with other theorists who defend the equal relevance of diverse perspectives on justice and the need to consider them with equal compassion during siting processes (Davy 1996 in particular). However, this conclusion seems to be more consistent with other established planning paradigms, particularly the collaborative planning paradigm (Healey 1997 and Healey 2003 in particular). Here, different worldviews are the communicative core of collaborative processes wherein actors aim at sharing the objectives, and the instruments through which achieving those objectives, of a spatial planning practice all bent toward societal justice.

I do not wish, however, to resolve this theoretical impasse by calling into play the collaborative planning paradigm, and the underlying constructivist epistemological position, *tout court*. I only wish to suggest that the paradigm offers a valuable reference framework for positioning the ethical discourse on hazardous facilities siting within the planning discourse. My understanding of "collaboration" relates to moral and normative rather than to restrictive epistemological positions, and implies an active role of citizens in achieving general planning objectives by departing from agreed reference values. As discussed in a previous contribution, such "active collaboration" among the actors involved in siting processes departs from an explicit definition of the respective moral rights and obligations (Basta et al. 2012), from the will to negotiate a shared notion of fairness that could accommodate the instances of the most disadvantaged while acknowledging the equal "spatial rights" of the whole, and from a compassionate acceptance of what such fairness entails in terms of residual sacrifices. The most important purpose of proposing such vision as a valid approach to the practical case of inventorying candidate sites for hazardous and risky facilities and planning the surrounding urban areas is the one of counter-balancing the techno-rational criteria of costs/benefits or risks/benefits with a *moral* criterion. While thinking at the strategic horizon of the planning practice, that is of one generation, and at the intergenerational horizon of the consequences of some of the risk-bearing technologies object of that practice, this approach should inform current and future siting processes with the highest possible disciplinary rigor.

In practical terms, this could imply opting not for the most economically advantageous or technologically smart solution but rather for the solution that represents the most even distribution of a spatially disadvantaged condition at the end of serving the objectives of the larger human community. In moral terms, I believe, this is something that people could be more prone to understand. As these last remarks address the future generation of "fair risk" planners, I wish to conclude with the simple yet most felt words of one of them: "the acceptability of risks does also depend on the distribution of those risks" (Bennebroek 2010).

Notes

1. I wish to thank my coeditor Stefano Moroni for having monitored the progresses of this chapter throughout its entire development. A warm thank goes to Stefan Koller, who patiently revised its final version and provided helpful critical remarks. Finally, I wish to thank all the participants of the workshop "The Ethics of the Built Environment" that led to the publication of this volume for their priceless contribution and progressively growing enthusiasm.
2. See, for instance, Rawls (1988, p. 257): "Provided due precautions are taken … we can in principle expand the list to include other goods…." "… If necessary the list of primary goods can in principle be expanded" (p. 257).

References

Arcuri A (2005) Governing the risk of ultra-hazardous activities: challenge for contemporary legal systems. Ph.D. thesis, Rotterdam Erasmus University, Rotterdam

Basta C (2009) Risk, territory and society: challenge for a joint European regulation. Ph.D. thesis, Delft University of Technology, Delft

Basta C (2011) Risk and spatial planning. In: Roeser S (ed in chief) et al. Handbook of risk theory. Springer, Berlin, pp 265–294

Basta C, van der Knaap W, Carsjens GJ (2012) Planning today the energy landscapes of tomorrow: from extended collaborative approaches to individual active planning. In: van Dobbelsteen A, Stremke S (eds) Sustainable energy landscapes: designing, planning and development. Taylor & Francis, London

Beck U (1992) Risk society: towards a new modernity. Sage Publication, London

Bennebroek B (2010) Ethical desirability of carbon capture and storage in Barendrecht. Student Essay, Delft University of Technology. Online at http://basbennebroek.nl/portfolio/ethical_desirability_ccs_barendrecht.pdf

Boholm A (2004) What are the new perspectives on siting controversies? J Risk Res 7(2):99–100

Boholm A, Lofsted R (2004) Facility siting: risk, power and identity in land use planning. Earthscan, London

Breunese JN, Remmelts G (2009) Inventory of potential locations for demonstration project CO2-storage, TNO-034-UT-2009-02024, The Netherlands

Brunsting S et al (2011) Stakeholder participation practices and onshore CCS: lessons from the Dutch CCS case Barendrecht. Energy Proced 4:6376–6383

Council Directive 96/82/EC on the control of major-accident hazards involving dangerous substances, OJ L 10/13

Davy B (1996) Fairness as compassion: towards a less unfair facility siting policy. Risk 7(2):99–108

Fainstein SS (2010) The just city. Cornell University Press, Ithaca

Gunder M (2008) Ideologies of certainty in a risky reality: beyond the hauntology of planning. Plann Theory 7(2):186–206

Hansson SO, Peterson M (2001) Rights, risks, and residual obligations. Risk Decis Policy 6:157–166

Harper T, Stein S (1992) The centrality of normative ethical theory to contemporary planning theory. J Plann Educ Res 11(2):105–116

Hayden Lesbirel S, Shaw D (2005) Managing conflict in facility siting. Edward Elgar, Cheltenam

Healey P (1997) Collaborative planning. Shaping places in fragmented societies. MacMillan Press, Houndmills/London

Healey P (2003) Collaborative planning in perspective. Plann Theory 2(2):101–123

Howe J, Langdon C (2002) Towards a reflexive planning theory. Plann Theory 1(3):209–225

Hudson B (2003) Justice in the risk society: challenging and re-affirming justice in late modernity. Sage, London

Huitema D (2002) Hazardous decisions. Hazardous waste siting in the UK, The Netherlands and Canada, institutions and discourses. Kluwer Academic Publisher, Dordrecht

Keller LR, Sarin RK (1995) Fair processes for societal decisions involving distributional inequalities. Risk Anal 15(1):49–59

Kuijper M (2011) Public acceptance challenges for onshore CO_2 storage in Barendrecht. Energy Procedia 4:6226–6233

Leiss W (2001) Ulrich Beck "Risk society" (1996) book review. Can J Sociol. Online at http://www.ualberta.ca/~cjscopy/articles/leiss.html

Linnerooth-Bayer J (2005) Fair strategies for siting hazardous waste facilities. In: Hayden Lesbirel S, Shaw D (eds) Managing conflict in facility siting. Edward Elgar, Cheltenam

Linnerooth-Bayer J, Fitzgerald KB (1996) Conflicting views on fair siting processes: evidence from Austria and the U.S. Risk 7(2):119–134

Linnerooth-Bayer J, Löfstedt RE (1996) Fairness and siting: introduction to a symposium. Risk 7(2):95–98

Moroni S (1994) Territorio e giustizia distributiva [Territory and distributive justice]. Angeli, Milano

Moroni S (1997) Etica e territorio [Ethics and land use]. Angeli, Milano

Owens S (2004a) Siting, sustainable development and social priorities. J Risk Res 7(2):101–114

Owens S (2004b) New agendas for appraisal: reflections on theory, practice, and research. Environ Plann A 36:1943–1959

Peterson M (2003) Risk, equality, and the priority view. Risk Decis Policy 8:17–23

Peterson M, Hansson SO (2004) On the application of right-based moral theories to siting controversies. J Risk Res 7(2):269–275

Rawls J (1971) A theory of justice. Harvard University Press, Harvard

Rawls J (1988) The priority of right and ideas of the good. Philos Public Aff 17(4):251–276

Roeser S (2006) The role of emotions in judging the moral acceptability of risks. Saf Sci 44:689–700

Roeser S (2010) Emotions and risky technologies. Springer, Berlin

Simmons P, Walker G (2004) Living with technological risk: industrial encroachment on sense of place. In: Boholm A, Lofsted R (eds) Facility siting: risk, power and identity in land use planning. Earthscan, London

Stein SM, Harper T (2005) Rawls's "Justice as fairness": a moral basis for contemporary planning theory. Plann Theory 4(2):147–172

Terwel BW et al. (2012) It's not only about safety: beliefs and attitudes of 811 local residents regarding a CCS project in Barendrecht. Int J Greenh Gas Control 9:41–51

VROM (2007) New energy for climate policy: the "clean and efficient" programme, online. Available at: http://www2.vrom.nl/docs/internationaal/New%20Energy%20for%20Climate%20Policy.pdf

Chapter 12
City Planning and Animals: Expanding Our Urban Compassion Footprint

Timothy Beatley and Marc Bekoff

12.1 Introduction: Little Attention to Animals in City Planning

For most of us living in cities or suburbs, there is relatively little recognition of or thinking about the other animals and life forms that occupy our planet, aside from the domesticated companion animals (pets) who share a special place in our households. We often forget that we coinhabit our landscapes and built environments with many "others," a rich and diverse array of animals and life, whose ethical and planning status is ambiguous to say the least. Little or no explicit attention has been given to animals in the planning literature, or in contemporary planning practice, despite the ubiquity of "animal questions", and the extent to which urban policy and urban development affect them. We argue here that this should change and provide many examples here of the ways in which the interests of animals can and should be integrated into planning their policy and practice. The contemporary values that underpin city and regional planning must we believe shift to include animals. The status and condition of animals, so impacted by planning policy at many levels, should become a legitimate and important topic of discussion within professional planning circles, as well as more generally in community planning processes and community engagement discussions.

T. Beatley (✉)
School of Architecture, University of Virginia, Charlottesville, VA, USA
e-mail: beatley@virginia.edu

M. Bekoff
Ecology and Evolutionary Biology, University of Colorado,
Boulder, CO, USA
e-mail: marc.bekoff@gmail.com

C. Basta and S. Moroni (eds.), *Ethics, Design and Planning of the Built Environment*, Urban and Landscape Perspectives 12, DOI 10.1007/978-94-007-5246-7_12,

In part this call is made more urgent by the growing scientific literature and research that shows compellingly that animals exhibit complex emotional lives and a level of moral behavior perhaps surprising to most. Bekoff's research (2007a, b, 2010, 2013a, b), and others, increasingly paints a picture of the animal world, where cooperation, empathy, justice, and fair play can be seen not only in cetaceans, primates, and elephants but also in mice, chickens, and rats (see especially Bekoff and Pierce 2009; Pierce and Bekoff 2012 for a review of this research and extensive references). Animals share more in common with *Homo sapiens* than we commonly accept, and we must begin to take them seriously as important members of our planning community.

There are, of course, many important (and different) environmental values and arguments that can serve to underpin explicit support for animals in cities: aesthetic and fascination value, the enjoyment and pleasure they provide urban populations (when they see them, hear, know of their existence), the understanding that humans have coevolved and require contact with other forms of life to be happy, productive and to live fully meaningful lives, and of course the belief that they hold intrinsic worth, irrespective of the instrumental values they might provide to urbanites (e.g., see Beatley 1994).

We become more aware of other animals when there are conflicts, of course: growth of deer populations in suburban settings, nesting turkey vultures who are perceived as nuisances, and increasingly the expansion of coyotes, a new presence in many urban environments. But these are, of course, only the most obvious examples of the nature around us, and its diversity even in cities is astounding, from the complex lichen on tips of trees to the millions of migratory birds moving through the city, to the subterranean invertebrates and aquatic species that inhabit spaces that are less visible, but quite proximate to where larger human populations reside. For the most part, we see little connection and form few bonds with this immense and fascinating biodiversity, and little reason to exercise more than casual attention in resolving occasional human-animal conflicts that emerge.

How we treat these "others" becomes a litmus test for our larger ethical sensibilities, and in many ways how we treat other human beings. And aside from our pets, we don't tend to treat them very well. A recent example from California is illustrative: in response to a report that a coyote nipped the toes of a napping visitor to Griffith Park, in Los Angeles, the entire group of seven coyotes was quickly killed by wildlife officials, even though the threat in this case, according to most coyote experts, was small to nil and need not have involved killing all of the coyotes in this urban pack. One of us (Marc Bekoff) consulted on this situation and suggested that there was no reason at all to kill the coyotes. Marc has also worked hard to get people in and around Denver, Colorado, to appreciate the presence of highly adaptable and intelligent coyotes (as a scientist who has studied coyotes for decades and as an advisor to Project Coyote (http://www.projectcoyote.org/)) and to understand that we have redecorated their homes and we cannot blame them for returning to what was theirs in the first place. Coexistence and respect must be our main goals because we cannot continue to ignore nature (Bekoff 2013a) and continue to abuse the very animals who have drawn us to where we live and recreate.

12.2 Need for an Animal-Inclusive Vision of City Planning

Our response to the growing presence of urban animals is often one of indifference at best, callousness at worst. There are often trade-offs and difficult decisions, to be sure, for instance, when a needed building or infrastructure adversely affects wildlife, but too often and too easily the animal interest is trumped or considered unimportant. An important but largely unaddressed question for planners becomes: How do we design places and lifestyles that are respectful and compassionate toward the other animals? How do we build corridors of compassion and coexistence?

Caring about and planning for the inclusion of the "others" seem very good goals for city planning, and finding new ways to curtail the huge human impact on global nature (some estimates predicts that global warming could cause the extinction of nearly 40% of current species by 2050) and to make room for animals in our cities (e.g., Block et al. 2001). There are many potential steps that could be taken, such as planting natural landscaping around our homes and buildings, to adopting bird-friendly design standards, as some cities, like Toronto, have done.

Cities could further expand corridors of protected greenspace and design them to allow dispersal, movement, and adaptation in response to growing numbers of people and changing climate. Some cities, like Brisbane, have developed plans designed to connect parks and greenspaces and to provide connections and corridors that will help species adapt in both the short and long term. This city has been installing wildlife movement structures that allow animals to cross over or above roadways (see Beatley and Newman 2009, for more detailed discussion). The animals and nature around us in cities and suburbs offer the possibility for wonder and fascination and contact with wildness and the natural world that is nearby. Nature is not "away"; it is "here," and as the planet becomes increasingly urban, finding ways to accommodate and coinhabit cities with other forms of life will become an even greater challenge.

And of course the animals around us offer the potential to improve our lives in many ways. They are wondrous often in their biology and life cycles, as well as the beauty and meaning they can add to the places in which we live. Jennifer Wolch, dean of the UC Berkeley School of Environmental Design, has written of the concept of Zoöpolis, understanding cities and communities as places where animals can co-occupy space, one important result being the "re-enchanting" of our cities (Wolch 1996; Seymour and Jennifer 2009). We like this idea very much, as a fuller appreciation of and care for the animals and nature around us is really about imbuing cities and suburbs with a new meaning—as places that are profoundly shared by a fascinating and wondrous subset of the planet's biological diversity.

12.3 The Value and Importance of Connections with Animals

There is considerable evidence that urbanites value and appreciate animals and wildness and understand that their presence is life enhancing and improves quality of life in cities. One dramatic example of this can be seen in Austin, Texas, where despite a rocky start, this city has now developed a love affair with its Mexican free-tailed bats who occupy the crevices of the Congress Avenue Bridge in downtown Austin. Thanks in large part to the leadership and advocacy of Bat Conservation International (which moved its offices to Austin), the city has gone from fearing the bats to celebrating them, now as a significant tourist attraction and economic engine for the city. And now the Texas Highway Department is even designing new bridges to accommodate bats, a major shift in the direction of coexistence.

Attending one of the bat "emergences" on a hot August night provides a glimpse into how important these bats have become and how fascinated people are with them. Even several hours before dark, families with coolers and blankets start to arrive, and a buzz of anticipation builds. Eventually the entire hillside east and south of the bridge is covered with people. On this night, a local band plays for the crowd (the Keep Austin Loud Project!), a kind of warm-up act, and as the sky darkens, as many as five tour boats jostle for the best positions to see the emergence. Eerily, people begin to line up along the east railing of the bridge, and a line of human bodies is silhouetted against the Austin sky.

Most heartening are the young children, many sitting spellbound, in the very front, exuding a kind of wondrous anticipation. We interview some of them and get a deeper sense of how drawn they seem to be to these creatures. Perhaps because they've learned about bats in school (several told us this), or because they have grown up in Texas hearing about the Austin bats, there's no fear or revulsion. Quite the contrary: it's an intense interest and fascination that, listening to them tell us why the bats are cool, I can't help but ponder why we tend to lose this perspective as we grow older.

It has been estimated by Bat Conservation International that some 100,000 people come to the bridge to see the bats each year, generating $10 million in ecotourism revenue. With 1.5 million bats, the Congress Avenue Bridge is believed to be the "world's largest urban bat colony." While originally viewed by local officials as a health threat and nuisance, Austinites are clearly proud of the bats and view their presence as something very special about the city. Each summer day, as evening approaches, thousands of residents and visitors line up on the bridge and surrounding areas to watch the bats' "emergence." It is a major daily event in that city! (see Beatley 2008). Austin's affection for bats is in evidence in other ways as well, including by naming its local hockey after them (the "Austin Ice Bats"). A bat observation deck and viewing area have been built by the Austin Statesman, the local newspaper (they actually call it the "Statesmen Bat Observation Center"). On any summer evening, there are several dinner boats plying waters of around the bridge, offering bat-watching dinner cruises up and down the Colorado. Every labor day weekend is held one of the city's most popular public event, the Bat Fest.

12.4 New Ways to Include Animals in City Planning

12.4.1 *Creative Strategies for Urban Coexistence*

Whether it is the sea lions at Pier 99 in San Francisco, or the bears and moose who inhabit the city of Anchorage, or the coyotes in the Chicago metro area or in and around Boulder, Colorado, urbanites like seeing and experiencing wild nature close by, and it significantly enhances quality of life in these places. This is not to deny that there may be some element of danger and cities need to proactively take steps to plan for humane and effective coexistence. There are alternatives to the scenario that unfolded in Griffith Park, and models of coexistence planning and actions that can be replicated. Cities could also adopt new educational and urban wildlife management efforts and protocols that reflect a new care for and concern about animals and which take a proactive approach to coexistence.

One of the most impressive and effective is the Vancouver's Co-Existing with Coyotes (CWC) Program, in existence since 2001. It emphasizes a combination of education and awareness raising and nonlethal response to coyote-human conflicts. Run by the nonprofit Stanley Park Ecology Society, the program uses the "two-pronged approach": short emergency response and long-term education (Worcester and Boelens 2007). The program maintains a coyote hot line and is able to respond effectively and with nonlethal means (noisemakers, for instance; the program shows how to make devices on its website). Long-term education includes visiting elementary schools, teaching "Coyotes 101," and helping students to learn how to identify coyotes, what steps to take in coexistence (no human feeding), and how to stay safe should a coyote approach. The program also conducts interpretive coyote walks through neighborhoods throughout the city. The CWC program is an exemplary example of how, with planning and with a spirit of coexistence, conflicts can be avoided and animals treated compassionately without resorting to lethal force while at the same time imparting a respectful sense of the wildness and value that coyotes can bring to urban life.

Many other common conflicts between wild animals and urban and suburban communities can be addressed through more humane means and techniques. New ways of managing perceived-to-be-nuisance resident populations of Canada geese, for instance, emphasize nonfatal methods such as public education and restrictions to feeding and use of sheep dogs. GeesePeace, for instance, is one effective new set of methods developed in suburban Virginia to address more humanely the year-round presence of Canada geese (see PeaceGeese, undated). Humane coexistence with other forms of life suggest a commitment as well to more effectively address homeowner and building owner treatment of animals (e.g., species such as bats in chimneys and raccoons in attics) that are routinely killed, often cruelly, when avoidance, exclusion, and sometimes relocation are possible. Support for local wildlife rehabilitation and care facilities would also be suggested; animals were truly taken into account in municipal policy and planning.

Domestic animals are, of course, a major presence in cities and local communities, and here there is an equally important set of policy and planning questions, again often absent from the usual planning agenda. Cities and counties often operate animal shelters and animal control agencies, raising significant issues about how sentient animals are treated. Many shelters have now shifted to a policy of "no kill," no longer euthanizing loss or unwanted domestic pets. With the rise of interest in urban agriculture and with many localities now modifying their zoning codes to permit farm animals in residential areas, the issues of their human treatment are important as well (e.g., consider the Seattle League for Goat Justice). As well, in many cities, urbanites are also seriously questioning the ethical aspects of their diets, including the impact on animals. Procurement policies in some cities has changed, for instance, to give preference to the purchase of eggs from free-range chickens, as Metro Vancouver, the regional planning agency there, has recently done (e.g., Vancouver Humane Society 2009).

12.4.2 Elevating the Status and Treatment of Animals in Cities

Cities are also home to various entertainment venues and activities that also raise serious questions about the status and treatment of animals, from zoos, to circuses, to rodeos. The City of Vancouver recently banned rodeos on the grounds of cruelty to animals (e.g., especially events such as calf roping), and many other localities in North America have taken or considered similar steps. The recent case of the Mirage Hotel in Las Vegas, importing two dolphins, with a NOAA permit allowing them to do so, is an especially egregious recent example of inhumane treatment, this following the death of many other dolphins at this hotel. The National Marine Fisheries Service, within NOAA, issued the dolphin permit for Mirage despite protracted opposition by groups such as Jean-Michel Cousteau's Ocean Futures Society, as well as Born Free USA and the World Society for the Protection of Animals (WSPA). Cousteau makes the case eloquently: "The more we learn about dolphins, the more we must admit they are our counterparts—intelligent, social, self-aware, capable of complex relationships, emotions, and learning. To consign them to a place like the Mirage hotel, with its 75 percent mortality rate for dolphins, and solely for our entertainment, is to impose a death sentence on innocents…In addition, the display of marine mammals for commercial gain does not represent the values we should be passing on to future generations" (Born Free 2009).

Animals and nature also inspire us in important ways (Bekoff 2010, 2013a, b), and we are increasingly learning much from them to solve contemporary problems. Janine Benyus has made a compelling argument about *biomimicry* and the many hidden lessons to be learned from nature (Benyus 2002). Nature reflects 3.8 billion years of research and development. But to learn from and be inspired by animals and nature, they need to be close by, and we need to appreciate how easy it is to have access to the amazing fauna and flora that live in our environs.

12.4.3 *Practices and Policies That Make Cities More Hospitable to Animals*

Planning practices and policies could be significantly adjusted to reduce the impact on animals. Urban development codes and design guidelines could easily be modified in many ways to make buildings and urban landscapes more hospitable to animals and other nature. New green areas and habitat can be found through the installation of green rooftops and green walls, by encouraging the planting of native vegetation around homes and buildings and reenvisioning the many leftover spaces in cities (from median strips to alleyways) as opportunities to support animals and nature.

Buildings in cities could be designed and redesigned to give more attention to the animals who come in contact with them. Notably many birds are harmed and killed by glass and lighting designs of high-rise buildings in cities. In cities like Chicago and Toronto, millions of migratory birds move through these cities at key times of the year. Toronto, perhaps more than any other city, has taken steps to lessen the impact of building on birds by developing a set of bird-friendly development guidelines and developing a program of recognizing developers and building owners who go above and beyond in designing their structures. In Toronto, as well as other cities like Chicago, "lights-out" campaigns have been underway to encourage building owners to turn off nighttime lights at key times of bird migration.

Recently, the Toronto environmental organizations Ecojustice and Ontario Nature have brought new attention to this problem by bringing legal actions against the owners of a complex of high-rise buildings, Consilium Place, found to be particularly dangerous to birds. Another local nonprofit, FLAP (Fatal Light Awareness Program), has documented very high bird mortality in response to the reflective glass of this complex, estimating the complex alone is responsible for some 7,000 dead and injured birds per year (FLAP, undated). The legal action, taken under Ontario's Environmental Protection Act as well as the Ontario Society for the Prevention of Cruelty to Animals Act, is in response to the unwillingness of the building owners to take fairly easy steps to prevent the carnage. That this issue is one of compassion for animals is clear when one considers how these birds die when they hit these building facades. In the words of one of the lawyers in the case: "Most of these birds die of traumatic injuries such as fractured skulls or broken backs" (Javed 2010).

12.4.4 *Reducing the Impacts of Urbanization and Development*

Reducing the spatial footprint of urban and suburban development is another important planning implication. Greater concern for animals and nature gives further support for curtailing sprawl, as western cities like Denver seem largely able to ignore impacts, for instance, on black-tailed prairie dogs, whose habitat has shrunk dramatically over time.

Animals, with the exception of federally or state-listed endangered or threatened species, receive little attention in the community planning and development process. This should change, and an animal-considerate community planning approach might look quite different. At a minimum, we must modify our environmental impact and development review processes and mechanism to better account for impacts on animals and animal communities. The loss of an oak forest in the process of building a new suburban shopping center take little account of the sentient animal life killed or displaced in the process, and indeed we have few analytic tool or methods to conduct such assessment, again because of largely indifferent view about animals. Few, if any, contemporary community plans include discussion of animals and their welfare, suggesting the need to fully and squarely include animals in the community of life for which we design and plan.

Another kind of detrimental urbanization occurs in waterways and ocean environments in the form of boat traffic and noise pollution. By one estimate, endangered North Atlantic right whales, known as the urban whales because they inhabit zones the heavily trafficked eastern US seaboard, are able to hear much less (only about 10%) than what they heard just a hundred years ago (Kraus and Rosalind 2007). The "acoustic smog" of oceans has increased dramatically from anthropogenic causes, and it is believed has significant implications for reproduction and long-term survival for cetaceans like the right whale, but there are other threats including entanglement in fishing lines and, most importantly, boat strikes. Opportunities exist here as well to plan, manage, and regulate with animals in mind. Evidence suggests that shifting shipping lanes can reduce boat strikes, and a recent NOAA rule now requires ships of a certain size (65 ft or greater) to reduce their speed to 10 nautical miles per hour in designated zones, seasonal management areas (SMAs). Preventing the death of these long-lived mammals must be a planning priority, and to be adopted and implemented often requires overcoming industry opposition and fears of negative economic impacts from such restrictions. Similar success can be seen in reducing boat strikes of manatees in Florida, through boat speed reduction zones.

The possibilities for protecting a wide variety of marine life have been given a lift in recent years with the new importance given to ocean and marine planning. A number of US state coastal management programs now include ocean management elements, and some regional planning agencies, such as the Cape Cod Commission, have now extended their planning jurisdiction well beyond the usual terrestrial environment (e.g., see Cape Cod Commission, undated).

12.4.5 *Making the Presence of Animals More Visible in Cities*

We are not likely to care for or about the life forms we cannot see, so finding new ways to make the animals more visible would also help. Aquatic and marine creatures represent a special challenge in this regard, and as Sylvia Earle, National Geographic oceanographer and marine explorer, eloquently notes, there is much biological diversity in the ocean (and in deep water) well beyond the usual attention of our terrestrial human world (Earle 2009). Recent efforts to track and monitor the movement of large ocean predators, many whose existence, including bluefin tuna, sea turtles, and whales, are yielding new insights about the biology and life cycles of these animals and, when their movements are mapped and overlaid, provide some helpful ways for distantly remote urban population to perceive and understand them.

We should also explore other creative ways to make animals in our cities and communities visible. These might include new ways of mapping the vertical diversity that exists in cities and finding ways to record, for instance, contrails of birds flying through our neighborhoods. Making camera traps available to neighborhoods interested in better (and more viscerally) understanding the nighttime animal life and equipping new urbanites and suburbanites with ecological owners' manuals that describe the fauna (and flora) likely to be encountered (and to be watched out for) would also help. Expanding our *compassion footprint* (Bekoff 2010) and "rewilding our hearts" (Bekoff 2013b) in cities may require other things of us, including time spent volunteering in urban habitat restoration projects and in no-kill animal shelters, helping to find homes for unwanted domestic pets, and in many other ways expanding the humane treatment of animals. The compassionate conservation movement (www.compassionateconservation.org; http://www.bornfree.org.uk/comp/compconsymp2010.html) is dedicated to achieving peaceful coexistence between human and nonhuman animals (Bekoff 2013a, b).

12.4.6 Looking Beyond City Borders

Compassionate and biophilic cities will also look beyond their borders to understand how their patterns of consumption and resource use impacts species and nature around the world (Beatley 2010). We know that the ecological footprint of a major city is tremendous in size and that supply lines for food, materials, wood, and energy are lengthy and often severely impacting on nature. Consumption of tropical wood by North American cities, for instance, is substantial, with direct impacts on the animals dependent on these habitats for long-term survival. Many cities are beginning to better understand these extra-local impacts on nature and take steps to curtail them. The City of New York, for instance, spends about $1 million per year on wood harvested in Brazil. It has recently made the decision to immediately reduce consumption by 20% and has developed a longer-term plan to further reduce its tropical wood consumption in the future. In Western Australia, the Perth Zoo has been leading a campaign to label products with palm oil and to raise awareness about the impacts of palm oil plantations on the plight of orangutans in Indonesia (Perth Zoo, undated).

Urban residents and city leaders can certainly have a significant impact in expressing care for our planet's animals, even creatures hundreds or thousands of miles away. When Tim was researching conservation success stories in Australia, he discovered the campaign to save Ningaloo Reef, a pristine fringing reef system in Western Australia, home to great biodiversity, notably spectacular whale sharks, but threatened by a large coastal development. Residents of Perth rallied to support the reef and were actually able to stop the development, even though most had never visited reef or seen the whale sharks and would likely never visit in the future (Beatley and Newman 2009). More recently, Australian port cities (e.g., Fremantle) have denied access to Japanese whaling vessels, in a clear demonstration of care and concern about whales. In a variety of ways, larger and small, cities can demonstrate their commitment to care and compassion of planetary life. Cities could also help support—with money, technical support, or volunteer labor—conservation programs and projects in other parts of the world, in part a recognition of the need to offset or compensate for the large ecological impacts associated with their lifestyle and consumption.

12.5 Concluding Thoughts: City Planning with Animals in Mind

Ultimately, there will be many good reasons, including economic, to incorporate animals more explicitly into planning practice and to give them more consideration in the theory and literature of planning (e.g., consider the tourism dollars generated from whale watching, for instance, and from watching the Mexican free-tailed bats in Austin). Protecting animals in turn serves to protect larger ecosystems and the ecological services, from retention and moderation of stormwater runoff to sequestering carbon, they provide. Animals are also a significant and important part of what makes a place or community distinctive, and evidence suggests that, even with a degree of danger or inconvenience, urban residents appreciate the value of these coinhabitants. A survey of attitudes about wildlife in Anchorage, Alaska, found, for instance, that while residents understood that moose in their city created certain problems, this wildlife also serves to make life there "interesting and special" (Alaska Department of Fish and Game 1999). Indeed, in large part, it is animals and nature that do much to define the special qualities of a place.

Perhaps most importantly, animals present the possibilities of profound wonder and wildness in the midst of urban and suburban grayness and banality. As with other aspects of nature and the natural world, direct access and exposure to animals and other forms of life have the potential to make us happier and more productive (Bekoff 2010, 2013b). And there are important ethical reasons why we must do a better job taking the interest of animals into account in city planning: we owe it to our fellow co-travelers—to acknowledge their inherent right to exist and in turn the ethical obligations we have to ensure that their survival and welfare are adequately taken into account in plans, policies, and decisions in communities large and small. The new research and emerging consensus about the moral and psychological complexities of animals further strengthen these ethical claims and, while not explicitly mentioned in our professional codes of ethics, suggest that planners have an ethical duty to plan for and humanely treat the many "others" with which we share this delicate world.

References

Alaska Department of Fish and Game (1999) Living with wildlife in anchorage. Found at: http://www.wc.adfg.state.ak.us/index.cfm?adfg=planning.anchorage5#values. Accessed 02 Aug 2010

Beatley T (1994) Ethical land use. Johns Hopkins University, Baltimore

Beatley T (2008) Toward biophilic cities. In: Kellert SR, Heerwagen JH, Mador M (eds) Biophilic design. Wiley, New York

Beatley T (2010) Biophilic cities. Island Press, Washington, DC

Beatley T, Newman P (2009) Green urbanism down under. Island Press, Washington, DC

Bekoff M (2007a) The emotional lives of animals. New World Library, Novato

Bekoff M (2007b) Animal matters. Shambhala, Boston

Bekoff M (2010) The animal manifesto: six reasons for expanding our compassion footprint. New World Library, Novato

Bekoff M (ed) (2013a) Ignoring nature no more: the case for compassionate conservation. University of Chicago Press, Chicago

Bekoff M (2013b) Rewilding our hearts. New World Library, Novato

Bekoff M, Pierce J (2009) Wild justice: the moral lives of animals. University of Chicago Press, Chicago

Benyus J (2002) Biomimicry: innovation inspired by nature. Harper Perennial, New York

Block B et al (2001) Migratory movements, depth preferences, and thermal biology of Atlantic bluefin tuna. Science 293:1310–1314

Born Free USA (2009) "Cousteau calls dolphin care at Las Vegas hotel a Mirage," media release. Found at: http://www.bornfreeusa.org/press.php?p=2078&more=1. Accessed 02 Aug 2010

Cape Cod Commission (undated) Coastal and marine resources. Found at: http://www.capecodcommission.org/coastal/regulation.htm. Accessed 02 Aug 2010

Earle SA (2009) The world is blue: how our fate and the ocean's are one. National Geographic Society, Washington, DC

Javed N (2010) Mirrored towers a fatal attraction to birds. The Toronto Star, 08 March

Kraus SD, Rosalind M (eds) (2007) The urban whale: North Atlantic right whales at the crossroads. Harvard University Press, Cambridge

FLAP (Fatal Light Awareness Program) (undated) About FLAP. Found at: http://www.flap.org/flap_home.htm. Accessed 02 Aug 2010

PeaceGeese (undated) The peacegeese solution. Found at: http://www.geesepeace.org/. Accessed 02 Aug 2010

Perth Zoo (Australia) (undated) Don't palm us off campaign. Found at: http://www.perthzoo.wa.gov.au/Get-Involved/Dont-Palm-Us-Off-campaign/. Accessed 02 Aug 2010

Pierce J, Bekoff M (2012) Wild justice redux: what we know about social justice in animals and why it matters. Soc Justice Res 25:122–139

Seymour M, Jennifer W (2009) Toward zoopolis: innovation and contradiction in a conservation community. J Urban 2(3):215–236

Vancouver Humane Society (2009) Annual report. Found at: http://www.vancouverhumanesociety.bc.ca/downloads/Annual%20Report%202009.pdf

Wolch J (1996) Zoopolis. Capital Nat Soc 7(2):21–47

Worcester R, Boelens R (2007) The co-existing with coyotes program in Vancouver, BC. In: Nolte DL, Arjo WM, Stalman DH (eds) Proceedings of the 12th wildlife damage management conference. National Wildlife Research Center, Fort Collins

Afterword: Ethical Problems of Contemporary Cities

Stefano Moroni

Our earth-transforming and landscape-creating capacities are basic qualities of our human nature: we are geographic beings (Sack 2003).

We human beings are the only city-building creatures in the world. The hives of social insects are fundamentally different in how they develop, what they do, and their potentialities (Jacobs 1961/1993).

Humans have used technology to transform the … world. … Whatever ethics we adopt will have to enable us to flourish in a technologically transformed world (Gunn 1998).

1 Introduction: Ethics for Contemporary Cities

1.1 The City as Living Environment

In Europe, more than 75 % of people live in cities (Le Galès 2002). Every month, five million more people live in the cities of the developing world (Glaeser 2011). Cities are wealth creators: over 80 % for developed nations (Landry 2008),[1] and they are cauldrons of cultural innovation. Despite all the hype over telecommunications and globalisation, cities are actually more important than before; the clustering force of cities is still fundamental (Florida 2008).[2] Cities have always been the heart of civilisation, but now, for the first time in our history, they have become the universal environment of human society, the "universal medium of people's lives on earth" (Schneider 2003).[3] It therefore becomes ever more crucial to question ourselves about what sort of ethical approach can be applied to today's cities.[4]

S. Moroni (✉)
Department of Architecture and Urban Studies, Milan Politecnico, Milan, Italy
e-mail: stefano.moroni@polimi.it; smoroni@aznet.it

C. Basta and S. Moroni (eds.), *Ethics, Design and Planning of the Built Environment*, Urban and Landscape Perspectives 12, DOI 10.1007/978-94-007-5246-7,

The kind of ethics that can be applied to the contemporary city involves a set of principles and rules of conduct that ensure the safeguarding not so much of the built environment *in itself* as the people living in it and their well-being.[5] The built environment can be a focus of ethical concern not per se – in its own right – but only in so far as this environment affects people or matters to them.

1.2 Taking Pluralism and Complexity Seriously

Two fundamental traits of contemporary cities are[6] (1) pluralism of the conceptions of the good and (2) the complexity of urban realities. These traits are the inescapable starting point for a reflection on a viable ethics for our cities: the idea is to search for an ethical approach that is not just abstract but grounded in our actual condition.

1.2.1 Pluralism

One "conception of the good" is an idea of what renders life pleasant and worth living – those things one regards as beneficial to oneself (Rasmussen and Den Uyl 2005). The different individual conceptions of good in our contemporary societies differ in terms of religious beliefs, cultural interests, lifestyles, etc. In other words, there are a great many forms of self-realisation (Larmore 1996). This pluralism of the conceptions of the good is not a mere contingent condition that may pass away; it is a permanent feature of our social environment (Rawls 2001). The plurality does not necessary depend on egoism but on the diversity of individuals' or groups' ideas of what constitutes the good life: even a society composed entirely of disinterested people, but with different altruistic ideals (regarding who and why is a worthy subject for our altruistic attentions), would be a pluralist society. In brief, pluralism depends on the multiplicity of preferences, interests, and desires of self-interested people and not on selfishness in the strictest sense.[7] Needless to say, modern cities have been the prime locus of concentration of different ideas about how life should be lived, of how one's time and resources should be employed. Clearly, this pluralistic nature is even more accentuated in the larger cities, starting with the so-called first-generation metropolises; this feature is endorsed even further by the current type of contemporary second-/third-generation metropolises, which are effectively utilised by different types of "populations": inhabitants, commuters, city users, etc. (Martinotti 1993). Moreover, it is worth noting how the recently growing phenomenon of immigration heightens the pluralist and multicultural nature of the city (Rogers 2001; Keith 2005; Syrett and Sepulveda 2012).

1.2.2 Complexity

Jane Jacobs (1961) was the first to clearly recognise the intrinsic complexity of urban systems.[8] Cities are highly complex systems indeed, in that they (1) have a very large number of components (individuals, activities, etc.) interacting in a polycentric way; they (2) present innumerable non-linear (nonadditive) interactions among those components, with many direct and indirect feedback loops; (3) they exhibit unintentionally emergent patterns; and they are (4) adaptive and dynamic (Portugali 1999; Batty 2005; Pumain 1998; Baynes 2009). The future of a city is therefore intrinsically undetermined: novelty and surprise are fundamental aspects of the urban process. Complexity is in this case clearly a property of the world itself, not simply something beheld by the observer. It amounts to more than mere complication.[9] It is important to stress that a complex urban order cannot be efficient in an engineeristic sense, exactly because it is conducive to experimentation and discovery. Cities are not efficient in this sense because they are incubators of new ideas and practices: in an urban environment where there is no perfect, static knowledge, innovation entails experimentation, trial and error, duplication, etc. (Ikeda 2007). To quote Jacobs (1969, p. 86): "I do not mean that cities are economically valuable in spite of their inefficiency and impracticability but rather because they are inefficient and impractical". One point is worth making before we conclude: many of those working in the field of land-use planning today seem to consider the theme of complexity as self-evident – actually, the theme has been scarcely fathomed in this field.[10]

1.3 Levels of Ethical Discourse

A viable ethics for the contemporary city has to recognise that pluralism of the conceptions of the goods and complexity are ineradicable elements of our social world. (Observe that accepting "pluralism of the conceptions of the good" does not necessarily mean adopting "pluralism/relativism of values", i.e. moral scepticism).[11] The acceptance of pluralism and complexity excludes certain ethical options, such as "communitarianism" and "conservatism", which tend – albeit in quite different ways – to underestimate the role and impact of pluralism among conceptions of the good and urban complexity. The acceptance of pluralism and complexity as inescapable conditions nevertheless leaves the way clear for some alternative types of ethical perspectives. It is not my intention here to defend a substantive ethical perspective (i.e. to defend a specific position among those compatible with pluralism and complexity) but merely to highlight certain general characteristics that it should have.

A reconsideration of an ethical perspective for contemporary cities should function on several levels and fronts and on two in particular: a discussion must be held first (1) on issues in *institutional ethics* (as regards local government action in particular) (Sect. 2) and second (2) on issues in *professional ethics* (regarding certain professions in particular, namely, land-use planners, architects, policy analysts) (Sect. 3). In both cases, the ethics do not perforce need to be mere recommendations but may also develop into something of a more operative nature, such as legislative meta-restrictions and meta-requirements for local government action and limits and obligations for professionals contained in codes of professional conduct

(Taylor 1992). As regards both institutional ethics and professional ethics, it is worth pointing out certain aspects of a more "procedural" than "spatial" (and "physical") nature, partly because the latter issues are extensively dealt with by other authors in this volume and partly because I believe that some procedural aspects are crucial in any event. In both cases – institutional and professional – the procedural aspects I will discuss here concern the question of *limits* (to politics in the first case and to knowledge in the second).

2 Institutional Ethics: Politics Within Limits

Among the various rules that the local government introduces, the fundamental ones are those that affect the use and transformation of land, spaces, and buildings; these include regulations for what may be built and transformed and where; what characteristics buildings and spaces must have and the standards they must comply with; what type of activities may be practised in certain buildings and places and on what conditions; which collective services must be guaranteed by private property developers; and which rules of inclusion and exclusion apply to public and private spaces.

This kind of building and planning rule – introduced at local government level – have an enormous impact on our lives and daily well-being and likewise on social network growth and economic activity (Beatley 1994; Ben-Joseph 2005; Needham 2006). This way of regulating the uses and transformation of buildings and land affects not only the overall configuration of the urban fabric itself, but it also impinges on the city's activities, influencing, for example, the cost and availability of housing; the chances for new businesses or retail projects, among others; the supply of services and infrastructure; and not least access to given places and activities.

The preliminary ethical question that arises concerns (1) the margins within which a local government may introduce regulations (Sect. 2.1) and (2) what goals it should pursue within the margins that are allowed (Sect. 2.2). As James Buchanan and Roger Congleton (2003, p. xx) write: "There has as yet come to be no widespread understanding that a non-monolithic, … non-omniscient politics requires an anchor in principle, less it remain subject to the capricious forces of rotating coalitional interests". The point is that we must rethink our rules with the basic aim of limiting the harm local governments can do, while preserving beneficial governmental activities (Brennan and Buchanan 2000). Democracy itself has to work within limiting conditions (Allan 2001).

2.1 Restrictions and Requirements

Contrary to what often happens today, I believe that the urgent need is to insist that when they introduce regulations, local governments adopt the following criteria: (1) *simplicity*, (2) *anti-bureaucratism*, (3) *impartiality*, (4) *stability*, and (5) *openness*.

2.1.1 Simplicity

In the first place, utmost simplicity must be applied to land-use regulations and building standards. We must dismiss complex rules and seek out simple rules for a complex world: in other words, "the proper response to more complex societies should be an ever greater reliance on simple legal rules" (Epstein 1995, p. 21). In many countries during the nineteenth century, important attempts were made to simplify the legal rules; since then, however, the law has been made more and more complex, accepting the wrong idea that law has to mirror the growing complexity of our societies (Kasper and Streit 1998). In this perspective, complex land-use and building issues quickly generated a host of equally complex rules.[12] Systems of complex rules – peculiar to the traditional and current land-use plans and building codes – are sets of rules that present the following features: "technicality", that is, the trait of those rules that require a high level of expertise to understand and apply them (this means that ordinary citizens are not able to directly know whether they are in compliance with the rules); "density", in reference to those rules that try to cover in minute details all aspects of certain actions or activities; "differentiation", which regards the plurality of different overlapping sources of law concerning a given situation; "indeterminacy", referring to the fact that to be able to decide whether a given action is illegal, it is necessary to deal with several factors provided for, none of which is decisive (the rules fail to give a clear yes/no answer) (Epstein 1995, pp. 23–25). Both theory and practice have demonstrated that this kind of complex rules do not work well because they overburden human cognition and inflict needless high compliance costs (Kasper and Streit 1998). To have simple rules – which avoid technicality, density, differentiation, and indeterminacy – is not a utopian dream but a workable alternative[13] and not least a real necessity: "The more complex the system, the greater the need for simple rules to achieve order" (Webster and Lai 2003, p. 211).

2.1.2 Anti-bureaucratism

Along the same lines, the second necessity is to drastically streamline bureaucracy. Over the last 30 years, the subdivision-approval process, for instance, has in many countries increased in its complication, as regards the number of agencies and committees involved, the number of standards and additional requirements, and the number of delays (Ben-Joseph 2005). Prolonged administrative-approval processes not only are prohibitive to developers (increasing their direct costs and their overall financial risk) but have also really negative consequences for the consumers (increasing, for instance, the final cost to the units buyers) (Ball 2010). According to a US research, for every additional month that was added to the completion date of the approval process, there was a 1/2 % increase in the final selling price of the unit (Ben-Joseph 2005). In brief, "administrative roadblocks add significantly to the cost of housing and truly constitute barriers to development" (Schill 2005, p. 12). Cities should therefore replace the current lengthy and uncertain permit processes with more "automatic" and faster ones.

2.1.3 Impartiality

Third, it is fundamental that local government sets rules that are as impersonal and as general as possible. This can happen by producing long-term rules that contain no reference to particular persons, objects, etc. In other words, rules must apply to an indeterminate number of future cases. Impartial rules provide the means for the realisation of the incommensurable different purposes of different individuals.

2.1.4 Stability

A fourth vital step is to ensure rules are consistent and reliable. Rules enable citizens to have dependable expectations – in general terms and over long periods of time – with regard to the actions of others and to the actions of the state itself. Stability is decisive if people are to be guided by law not only in their short-term decisions and actions but also in their long-term ones. It is difficult to know, abide by, and respect, rules that constantly change; if legal rules are continually subject to change, the information they provide becomes negligible and useless (Brennan and Buchanan 2000). On the other hand, stability improves reliability, with the consequence of facilitating human interaction. Clearly, the only rules that can remain stable are those that deal with general aspects of local urban reality and do not claim to control the details. In other words, it is due to the tendency to apply overly detailed and specific regulations that we have avoided or failed to ensure stability to land-use and building rules.

2.1.5 Openness

A fifth requirement is to allow far more leeway for experimentation. In this perspective, rules must guarantee that the actions of individuals are coordinated only as regards their "typical features" (i.e. their repeatable, time-independent, and situation-independent aspects), not as regards their "specific features" (i.e. their unrepeatable, time-dependent and situation-dependent aspects) (Moroni 2011). In other words, the rule framework should be "open", so as to allow individuals (citizens, developers, architects, designers) to respond to new circumstances through innovative action, as suggested by their particular knowledge of circumstances of time and place. In brief, the local government works best when it sets the rules of the game, not when it seeks to determine specific end-states. For instance, as Eran Ben-Joseph (2005, p. 109) observes: "Excessive street and right-of-way widths, rigid earthwork specifications, and overdesigned infrastructure systems are unfavorable to the introduction of site-sensitive solutions, and often impede cost reductions". And he continues: "Obviously there are many issues to tackle in shaping a new regulatory template for subdivisions. But none is more important than the realization that this new template must allow and promote a variety of housing styles and types of development design" (p. 115). In this perspective, building codes and standards should only be used as a baseline and not as a barrier to prevent experimentation and innovation.[14] It is important in particular to place more emphasis on "performance standards" than on "physical specifications".

In conclusion, observe how many of the previous recommendations are good also to minimise public officials' corruption. Some of the above-mentioned pre-requisites work as a "veil mechanism" that suppresses illegitimate, self-interested behavior on the part of public decision-makers by subjecting them to uncertainty about the exact distribution of costs and benefits that will result from their decisions (Vermeule 2007).

The point is not that land-use officials and developers are especially prone to corruption but that the traditional legal-administrative land-use systems intrinsically instigate action in that direction. As John Gardiner (1985, p. 122) observes: "Corruption can only occur when an official has an opportunity to use his or her authority in a way which would lead someone to want to pay for favorable treatment".

Indeed, it is where discretionality is highest – and where there is greater possibility of differentiating between the positions of single individuals (land owners, developers, and so forth) by way of public decisions – that we find increased levels of corruption. A study sponsored by the US National Institute of Law Enforcement and Criminal Justice found that land-use decisions "were particularly susceptible to corruption because of the significant financial losses and gains which are imposed as a consequence of [traditional] zoning"; other corruption incentives include "the confused treatment of zoning as both a legislative and administrative matter, the increasing complexity of land use procedures, and the lack of standards guiding [land-use] decisions" (Kmiec 1981, p. 45).

It should be strongly emphasised that corruption among public officials is not just wrong in itself but brings negative consequences, such as the loss of competition between urban developers, and that building permission is granted to unsuitable developers.

In brief, the problem of corruption is likewise a crucial element of an ethics of the built environment. And the fact that it is often overlooked is proof of how such "procedural" ethical questions are underestimated.

2.2 *Objectives*

2.2.1 A Just and Creative City

Regarding the objectives, I suggest that they pivot on establishing a "just and sustainable city" that is also "creative" (Florida 2005, 2007; Andersson et al. 2011). These two factors do not necessary pull in the opposite direction if we avoid too static and reductive an idea of justice – for example, the aforementioned communitarian and conservative outlooks. At any event, it is no easy task to reach a balance between the two aspects: tackling it involves proposing a specific, substantive ethical perspective (which must also contain some indications of a spatial nature).[15] But this is not the object of the present chapter.

Generally speaking, I will merely say that the fundamental question is not to attract a fluctuating predefined "creative class" but to favour the institutional and social conditions so that everyone can become creative, in a perpetually experimental urban environment that involves all. Richard Florida has recently claimed that only a simplistic reading of his earlier works has led people to believe that he was focusing only on one sector of society, namely, the creative class as an exclusive elite. Irrespective of the correct interpretation of his earlier works, it is interesting to note that Florida himself later kept emphasising that "Creativity is as biologically and intellectually innate a characteristic to all human beings as thought itself" (2005, p. 4). In other words: "Every single person is creative in some way" (p. 22).

2.2.2 People as the Ultimate Resource

The crucial issue is that resources do not exist in their own right, independently from us. They are not a fixed quantity – a stock – whose contours are predefined. Resources depend on human desires and perceptions, knowledge, and technology. Strictly speaking, resources are therefore not so much "discovered" as "invented" (Kirzner 2000; Bauer 2004). As Leila Kebir and Olivier Crevoiser (2008, pp. 49–50) write: traditional approaches "consider that resources exist independently of production… In this case, the resources are reified: they exist in their own right, independently of the relations among the players and independently of the production processes". The basic question to which this perspective seeks a response "is how to allocate the existing resources in an effective way, given an objective that is defined? The scarcity of the resources is presupposed". A different, more promising approach "considers the resources as being constructed, meaning they are not imposed once and for all, but are relative and evolutive". Innovation plays a fundamental role in this case: "What constitutes, or will constitute, a resource will depend not only on what is imposed at the outset and in the future, but also on the intentions and perceptions of the actors". Actually, the properties of any given "material" are irrelevant if we do not know how to take advantage of them. The point is that, in the city environment also, the human being is the *ultimate resource* (Simon 1996). In brief: "Cities have one crucial resource – their people" (Landry 2008, p. xii).

3 Professional Ethics: Knowledge Within Limits

3.1 Common Duties and Special Duties

Among the main imperatives that professionals (planners, architects, etc.) must conform with are: (1) respect for one's colleagues and loyalty towards clients; (2) the correct application of knowledge, techniques, and data; (3) to avoid deceit;

(4) to avoid conflicts of interest; (5) to refuse any type of payment or favour that would benefit one party over another; and (6) to pursue a high-quality outcome, while taking account of possible repercussions of a more general nature regarding their choices and actions.[16]

The first five imperatives – which correspond to the requirement to be honest, truthful, and fair in their actions – are of a more "procedural" nature and tend to be similar across different professions, such as planners and architects. The last is of a more "substantive" nature and tends to differ between professions. This last type of responsibility has been discussed in other chapters in the present volume.[17] The first five imperatives are usually considered the most self-evident, and yet certain features of today's city – for example, its intrinsic complexity – actually generate problems that concern these "obvious" five imperatives. Regarding the correct application of know-how (and being honest), I believe that a crucial point prompted by the typical features of the city listed above is the acknowledgement of our limitations. For instance, the limits to our ability to forecast events and outcomes.

3.2 *Intrinsic Limits*[18]

3.2.1 Specific Predictions and Pattern Predictions

One of the fundamental consequences of the complexity of systems such as the city is indeed our inability to make any "specific prediction" on our future; we can only venture "pattern predictions". A "specific prediction" is one capable of predicting certain discrete events with a sufficient degree of precision. As with any form of prediction, a specific prediction states only some – and never all – of the properties of a particular phenomenon, but it can narrow down (circumscribe) these properties and can do this in quantitative terms. Conversely, a "pattern prediction" does not predict particular events but only peculiarly wide classes of events or, better, broad patterns of events; it can only indicate what the "kind" of expected event is (Hayek 1967). The term pattern prediction indicates that we are able to make only a qualitative (rather than a quantitative, precise numerical) conditional prediction about the phenomenon at issue (Caldwell 2004). For instance, we know what general effect a change in housing demand will have on housing prices, but we cannot predict in detail what quantitative changes will occur.

3.2.2 Urban Models

Bearing all this in mind, at this point it is worth making some observations about urban models. In the field of urban studies, some formal models – that show greater sophistication than the traditional ones so far proposed – have recently been put forward to explain how cities function (Wilson and Bennett 1985; Dendrinos and Sonis 1990; Batty and Longley 1994; Batty 2005). As Michael Batty (2005, p. 516) notes at the end of his fascinating book on urban modelling: "Our models have attempted to extract the essence of dynamic processes generating urban development, but invariably this kind of abstraction focuses on generic outcomes rather than specific predictions". Furthermore: "Our models simply provide ways of thinking about cities" (p. 517); "these kinds of models inform but do not predict" (p. 518). In line with the discussion developed above, not only is this not a bad thing, but it is precisely how it *should* be. In this regard, the point is not so much that the models "are not the reality" – as every theory or model must perforce involve some level of abstraction – but that *different realities* require *different models* which can provide certain answers and not others.

3.2.3 The Practicability and Relevance of the Models

We must therefore recognise that proper models of urban complexity cannot avoid giving explanations of the principle and pattern predictions, not because they are "models" but because they are models of a specific complex reality, namely, the city. As Friedrich von Hayek (1967, p. 16) observes, referring to models that provide only explanations of the principle and pattern predictions: "Such models are valuable on their own, irrespective of their use for determining particular situations, and even where we know that we shall never have the information which would make this possible". And he continues: "The understanding of the general mechanism which produces patterns of a certain kind is not … a tool for specific predictions but important in its own right, and … it may provide important guides to action – or sometimes indications of the desirability of no action" (p. 40). We are accustomed to thinking that the knowledge generated by our models is useful in directly guiding our action; instead, the real usefulness of models of this kind is often to make plain what we cannot know or do. In my opinion, many complex models are useful inasmuch as they show up our *structural ignorance* (Moroni 2012).[19] It is worth noting how decision-makers often complain that certain models fail to support unequivocal choices, for example, when taking localised decisions – such as, a theme park X must (or must not) be sited in area C. But, as argued above, this fact does not so much denote the limit *of the models* as the incorrect idea that we can make specific land-use decisions on the basis of an unattainable knowledge of detail. As Nicholas Rescher (2009, p. 94) writes: "Questions whose resolution requires determining the outcome of contingent future events … are … not … answerable in a convincingly cogent way".

3.3 *Acknowledging Limits*

To conclude, it is the moral duty of the professional – the land-use planner, for instance – not to pretend that certain urban models are able to perform beyond a certain point. More generally, it is the moral duty of all professionals working in the contemporary city not to claim a thorough knowledge of (and ability to predict) the urban process beyond what is actually feasible. Such a vigilant and cautious approach is not always adopted, notably in the planning field.[20]

In general terms, we must recognise that constraints to perfect knowledge come not only from us but also from external conditions. They do not depend simply and solely on computational limits of our minds but on the structure of the world itself (Moroni 2012). Our limits in explaining and predicting are therefore both epistemically and ontologically grounded.[21]

4 Conclusions: An Ongoing Debate

The aim of this book has certainly not been to supply a univocal solution to the complex ethical problems of the contemporary city but rather to highlight how these issues are ever-present and require due attention; they affect different levels of action (institutional, professional, etc.) and impinge on diverse professional fields (planning, architecture, design, etc.).

This afterword in no way aimed to summarise all the book's elements but instead to add some extra elements of discussion to an ongoing debate that requires further inquiry. I conclude with the note of hope expressed by Jane Jacobs (1961/1993, p. 584): "Being human is itself difficult, and therefore all kinds of settlements (except dream cities) have problems. Big cities have difficulties in abundance ... But vital cities are not helpless to combat even the most difficult of problems... Vital cities have marvelous innate abilities for understanding, communicating, contriving and inventing what is required to combat their difficulties".

Notes

1. Between prosperity and urbanisation, there is a near-perfect correlation: "on average, as the share of a country's population that is urban rises by 10 %, the country's per capita output increases by 30 %" (Glaeser 2011, p. 7).
2. On this point, see also Clarke and Gaile (1998), Sassen (2000), Landry (2008), and Glaeser (2011).
3. "On a planet with vast amounts of space (all of humanity could fit in Texas – each of us with a personal townhouse), we choose cities" (Glaeser 2011, p. 1).
4. Ethics involves principles that regulate choices and actions, telling us what we ought or ought not to do. I am obviously speaking here about "normative ethics" (i.e. a search for action-guiding principles) and not about "descriptive ethics" (i.e. an attempt to discover what ethical criteria citizens, technicians, public officials, etc. actually accept). For this distinction, see, for instance, Cooper (1993). Interesting studies in descriptive ethics (with particular attention to land-use planners) are Vasu (1979), Howe and Kaufman (1981, 1985), Hendler (1991), and Howe (1994).
5. The question of non-human occupants of the city is omitted, as it does not impact on the central point stressed here.

6. This section is based on Moroni (2012).
7. Note that not even the market itself can be said to be driven solely by self-interest: in market processes, self-interest is indeed a crucial element, but "properly understood self-interest does not exclude altruistic motivation; it depends on purposefulness, but not on any selfishness of purpose. The point to be stressed is that it is one's own purposes which inspire one's actions ... One's purposes may be altruistic or otherwise... It is human dreams and goals which provide the motive force for market processes" (Kirzner 1992, p. 208).
8. Jacobs (1961) started asking what kind of problem cities pose. In order to find an answer, she distinguishes between three type of problems: (1) "problems of simplicity" (situations which presents two factors which are directly related to each other), (2) "problems of disorganised complexity" (problems presenting many factors interacting in multiple ways but with no stable coherent pattern of interrelation), and (3) "problems of organised complexity" (problems presenting many factors which are interrelated in an organic whole). She concluded that cities are without doubt problems in organised complexity (1961/1993, p. 564). On the contrary: "The theorists of conventional modern city planning have consistently mistaken cities as problems of simplicity and of disorganized complexity, and have tried to analyze and treat them thus" (Jacobs 1961/1993, p. 567).
9. As Atlan (1981, p. 186) writes: "We shall say a system appears complex when we do not know how to specify it completely although we know enough about it to recognize it [...]. In this respect, complexity must be distinguished from what we may call complication: The latter only expresses a high number of steps necessary to describe a system".
10. As aptly pointed out by Byrne (2003), Portugali (2008), and Innes and Booher (2010). See also Moroni (2012).
11. See Rawls (1993), Rescher (1993), Waldron (1993), Larmore (1996), and Connolly (2005). Pluralism of the conceptions of the good as interpreted here is the recognition of an inevitable disagreement about what constitutes a good life and the recognition that people tend to disagree about the nature of self-realisation. But this does not mean necessarily (Moroni 2004) that there exists (1) no value at all that overrides others (state neutrality towards conceptions of the good is itself a meta-value: Waldron 1993) or (2) no limit to freedom of choice (issues regarding the "good" – the way of life we prefer for ourselves, where the state must not enter – can be separated from issues regarding the "right" – the actions that create harm to others and that must be prohibited by the state: Rasmussen and Den Uyl 2005, pp. 22–28). In the normative perspective I accept here, the point is simply that the state cannot impose any comprehensive conception of the good life on the people. The state must therefore concentrate on defending the right of each person to pursue the conception of the good life that he or she prefers, without harming others, and on guaranteeing to everyone – in particular to the worst-offs – also certain "means to differing ends".
12. "Through the years, the design and layout of urban developments have become increasingly regulated. Professional and governmental bodies have developed standards for the built environment that dictate all aspects of the form and shape of urban ... communities. Obviously, development standards can assure a level of quality in performance as do those plans and construction standards designed to protect our health and safety. The problem arises when standards intended for health and safety overstep their bound and lose grounding in the objective measures of their benefit or break the connection with the original rationale for their existence. This disconnection has overtaken many standards and regulations today" (Ben-Joseph 2005, p. 2).
13. "The common perception is that it is idle at best to long for a return to the imagined simplicity of some past gilded age. Criticisms of legal complexity are often greeted with a shrug by those who view the proliferation of legal rules as an unavoidable necessity" (Epstein 1995, p. 23). "But the current situation is neither inevitable nor desirable" (p. 21).
14. Instead, traditional regulations and urban plans have generally limited the space for experimentation and innovation in constructions and settlements. In a national survey of American developers conducted by Ben-Joseph (2005, p. 105), developers made three

recurrent comments: (1) "regulatory agencies exceed their authority to practice social engineering, architecture, and micromanagement"; (2) subdivision codes "do not allow any flexibility"; and (3) city and county offices "are only interested in exactions and imposing regulations that make them appear more successful in protecting the community from the 'evil' developer that may be trying to be profitable". According to data gathered by Ben-Joseph, more than 70 % of American developers maintain that the main problem that they have to tackle in developing projects consists in the city-planning regulations, which are too detailed and invasive (along with the endless procedures and bureaucratic steps they involve). This is a problem that developers consider more serious than the finding of areas or financial resources. Not so much developers who want to built without rules, but developers who want to work with better rules.

15. On this score – and I agree with what other authors have pointed out elsewhere in this volume – I would urge that an ethical perspective for the contemporary city must also provide specific coverage for the spatial, physical aspects of the city (as cogently stressed by Lynch 1981); it is however important not to forget that the "institutional quality" comes before the "urban quality" and that the latter must be pursued within the framework of legitimacy guaranteed by the former (i.e. the "livability" of the institutions precedes the "livability" of the places and creates the conditions for the latter to come about).
16. I assume here that the professional practitioner (for instance, the professional planner) is neither a "mere technician", totally value-neutral, nor a "political activist", completely value-laden, but that he/she is located somewhere in the middle. Irrespective of the way in which different perspectives qualify this "middle position", there is nonetheless no discussion that, in a democratic society, professional practitioners (for instance, planners) are first of all "experts" (citizens with extensive, special knowledge and particular skills in a specific field) and surely not (elected) "decision-makers". For an interesting discussion on this point, see Mazza (1995).
17. For more on this last point, as regards the ethics of the planning profession, see Wachs (1985), Thomas and Healey (1991), and Hendler (1995); as regards architecture, see Wasserman et al. (2000), Spector (2001), Ray (2005), Owen (2009), and Taylor and Levine (2011).
18. This section is based on Moroni (2009).
19. "The issue of contingent ignorance – of what people are too lazy or too incompetent to find out about – does not hold much interest ... What matters from the theoretical point of view are those aspects of ignorance that betoken inherent limits to human knowledge" (Rescher 2009, p. 3). "It is important to heed the distinction between facts that nobody *does actually* know and facts that nobody *can possibly* know – between merely unknown facts and inherently unknowable ones. ... The really interesting issue, accordingly, relates not to what is not known to some ... Instead, the really interesting question relates to that which cannot be known at all" (p. 4).
20. As Portugali (2008, p. 250) rightly observes, "the structure of planning law, practice and administration, [is] [...] based on the (usually implicit) assumption that cities are essentially predictable entities; that given sufficient data and information, their future behavior is in essence predictable". See also Staley (2004, p. 273): the presumption underlying most recent planning law and practice in the United States "is that all the relevant factors for determining housing demand and supply, land availability, and the interrelationships between commercial, industrial, and residential land development are known and foreseeable".
21. Compare with Rescher (2009, p. 101): "In principle, there are both ontological and epistemological limits to predictive foreknowledge, and obstacles to successful prediction can reside either in the nature of things or in our own cognitive limitations. Ontological limits exist insofar as the future of the domain at issue is *developmentally open* – causally undetermined or underdetermined by the existing realities of the present and open to the development of wholly unprecedented patters owing to the contingencies of choice, chance, and chaos. Epistemological limits on prediction exist insofar as the future is *cognitively inaccessible* – either because we cannot secure the needed data, or because it is impossible for us to discover the operative laws, or even possibly because the requisite inferences and/or calculations involve complexities that outrun the reach of our capabilities".

References

Allan TRS (2001) Constitutional justice. Oxford University Press, Oxford
Andersson DE, Andersson AE, Mellander C (eds) (2011) Handbook of creative cities. Edward Elgar, Cheltenham
Atlan H (1981) Hierarchical self-organization in living systems. In: Zeleny M (ed) Autopoiesis. Elsevier, New York, pp 185–208
Ball M (2010) Planning delay and the responsiveness of English housing supply. Urban Stud 48(2):349–362
Batty M (2005) Cities and complexity. The MIT Press, Cambridge
Batty M, Longley P (1994) Fractal cities. Academic, London
Bauer PT (2004) From subsistence to exchange and other essays. Princeton University Press, Princeton
Baynes TM (2009) Complexity in urban development and management. J Ind Ecol 13(2): 214–227
Beatley T (1994) Ethical land use. The Johns Hopkins University Press, Baltimore
Ben-Joseph E (2005) On standards. In: Ben-Joseph E, Szold T (eds) Regulating place. Routledge, London, pp 1–14
Brennan G, Buchanan JM (2000) The reason of rules. Liberty Fund, Indianapolis
Buchanan JM, Congleton RD (2003) Politics by principle, not Interest. Liberty Fund, Indianapolis
Byrne D (2003) Complexity theory and planning theory: a necessary encounter. Plann Theory 2(3):171–178
Caldwell B (2004) Hayek's challenge. The University of Chicago Press, Chicago
Clarke SE, Gaile GL (1998) The work of cities. University of Minnesota Press, Minneapolis
Connolly WE (2005) Pluralism. Duke University Press, Durham
Cooper D (1993) Value pluralism and ethical choice. St. Martin's Press, New York
Dendrinos DS, Sonis M (1990) Chaos and socio-spatial dynamics. Springer, Berlin
Epstein RA (1995) Simple rules for a complex world. Harvard University Press, Cambridge, MA
Florida R (2005) Cities and the creative class. Routledge, London
Florida R (2007) The flight of the creative class. Collins, New York
Florida R (2008) Who's your city? Basic Books, New York
Gardiner JA (1985) Corruption and reform in land-use and building regulation. In: Wachs M (ed) Ethics in planning. CUPR, New Brunswick, pp 121–142
Glaeser R (2011) Triumph of the city. Macmillan, London
Gunn AG (1998) Rethinking communities: environmental ethics in an urbanized world. Environ Ethics 20:341–360
Hayek FA (1967) Studies in philosophy, politics and economics. Routledge, London
Hendler S (1991) Ethics in planning. The views of student and practitioners. J Plann Educ Res 10(2):99–105
Hendler S (ed) (1995) Planning ethics. CUPR, New Brunswick
Howe E (1994) Acting on ethics in city planning. CUPR, New Brunswick
Howe E, Kaufman J (1981) The values of contemporary American planners. J Am Plann Assoc 47(3):266–278
Howe E, Kaufman J (1985) The ethics of contemporary American planners. In: Wachs M (ed) Ethics in planning. CUPR, New Brunswick, pp 25–50
Ikeda S (2007) Urbanizing economics. Rev Austrian Econ 20(4):213–220
Innes JE, Booher DE (2010) Planning with complexity. Routledge, London
Jacobs J (1961) The death and life of great American cities. New York: Random House. New York: Modern Library, 1993
Jacobs J (1969) The economy of cities. Vintage Books, New York
Kasper W, Streit ME (1998) Institutional economics. Edward Elgar, Cheltenham
Kebir L, Crevoisier O (2008) Cultural resources and regional development. In: Cooke P, Lazzaretti L (eds) Creative cities, cultural clusters and local economic development. Edward Elgar, Cheltenham, pp 48–69

Keith M (2005) After the cosmopolitan? Routledge, London
Kirzner I (1992) The meaning of market process. Routledge, London
Kirzner I (2000) Entrepreneurship, entitlement, and economic justice. In: Vallentyne P, Steiner H (eds) Left-libertarianism and its critics. Palgrave, Basingstoke, pp 191–213
Kmiec DW (1981) Deregulating land use. Univ Pa Law Rev 130(1):28–130
Landry C (2008) The creative city. Earthscan, London
Larmore C (1996) The morals of modernity. Cambridge University Press, Cambridge
Le Galès P (2002) European cities. Oxford University Press, Oxford
Lynch K (1981) A theory of good city form. MIT Press, Cambridge, MA
Martinotti G (1993) Metropoli. Il Mulino, Bologna
Mazza L (1995) Technical knowledge, practical reason and the planner's responsibility. Town Plann Rev 66(4):389–409
Moroni S (2004) Towards a reconstruction of the public interest criterion. Plann Theory 3(2): 151–171
Moroni S (2009) Complexity, knowledge, and regulation. In: Rabino G, Caglioni M (eds) Planning, complexity and new ICT. Alinea, Firenze, pp 177–184
Moroni S (2011) Land-use regulation for the creative city. In: Andersson DE, Mellander C, Andersson A (eds) Handbook of creative cities. Edward Elgar, Aldershot, pp 342–364
Moroni S (2012) Why nomocracy: structural ignorance, radical pluralism and the role of relational rules. Prog Plann 77(2):46–59
Needham B (2006) Planning, law and economics. Routledge, London
Owen G (ed) (2009) Architecture, ethics and globalization. Routledge, London
Portugali J (1999) Self-organization and the city. Springer, Berlin
Portugali J (2008) Learning from paradoxes about prediction and planning in self-organizing cities. Plann Theory 7(3):248–262
Pumain D (1998) Urban research and complexity. In: Bertuglia CS, Bianchi G, Mela A (eds) The city and its sciences. Physica Verlag, Heidelberg, pp 323–361
Rasmussen DB, Den Uyl DJ (2005) Norms of liberty. Pennsylvania State University Press, University Park
Rawls J (1993) Political liberalism. Columbia University Press, New York
Rawls J (2001) Justice as fairness: a restatement. Harvard University Press, Cambridge, MA
Ray N (ed) (2005) Architecture and its ethical dilemmas. Taylor & Francis, London
Rescher N (1993) Pluralism. Clarendon, Oxford
Rescher N (2009) Ignorance. University of Pittsburgh Press, Pittsburgh
Rogers A (2001) Citizenship, multiculturalism and the European city. In: Bridge G, Watson S (eds) A companion to the city. Blackwell, Oxford, pp 282–291
Sack R (2003) A geographical guide to the real and the good. Routledge, London
Sassen S (2000) Cities in a world economy. Pine Forge Press, Thousand Oaks
Schill MH (2005) Regulation and housing development: what we know. Cityscape 8(1):5–19
Schneider KR (2003) On the nature of cities. Author Choice Press, New York
Simon JL (1996) The ultimate resource 2. Princeton University Press, Princeton
Spector T (2001) The ethical architect: the dilemma of contemporary practice. Princeton Architectural Press, New York
Staley SR (2004) Urban planning, smart growth, and economic calculation. Rev Austrian Econ 17(2):265–283
Syrett S, Sepulveda L (2012) Urban governance and economic development in the diverse city. Eur Urban Reg Stud 19(3):238–253
Taylor N (1992) Professional ethics in town planning. What is a code of professional conduct for? Town Plann Rev 63(3):227–241
Taylor WM, Levine MP (eds) (2011) Prospects for an ethics of architecture. Routledge, London
Thomas H, Healey P (eds) (1991) Dilemmas of planning practice. Avebury, Aldershot
Vasu ML (1979) Politics and planning. A national study of American planners. The University of North Carolina Press, Chapel Hill
Vermeule A (2007) Mechanisms of democracy. Institutional design writ small. Oxford University Press, Oxford

Wachs M (ed) (1985) Ethics in planning. CUPR, New Brunswick
Waldron J (1993) Liberal rights. Cambridge University Press, Cambridge
Wasserman B, Sullivan P, Palermo G (2000) Ethics and the practice of architecture. Wiley, New York
Webster C, Lai LWC (2003) Property rights, planning and markets. Edward Elgar, Cheltenham
Wilson AG, Bennett RJ (1985) Mathematical methods in human geography and planning. Wiley, Chichester

Author Index

C. Basta and S. Moroni (eds.), *Ethics, Design and Planning of the Built Environment*,
Urban and Landscape Perspectives 12, DOI 10.1007/978-94-007-5246-7,

Subject Index

C. Basta and S. Moroni (eds.), *Ethics, Design and Planning of the Built Environment*, Urban and Landscape Perspectives 12, DOI 10.1007/978-94-007-5246-7,

MIX
Papier aus verantwortungsvollen Quellen
Paper from responsible sources
FSC® C105338

If you have any concerns about our products,
you can contact us on
ProductSafety@springernature.com

In case Publisher is established outside the EU,
the EU authorized representative is:
Springer Nature Customer Service Center GmbH
Europaplatz 3, 69115 Heidelberg, Germany

Printed by Libri Plureos GmbH
in Hamburg, Germany